SpringerBriefs in Applied Sciences and Technology

Computational Intelligence

Series Editor

Janusz Kacprzyk, Systems Research Institute, Polish Academy of Sciences, Warsaw, Poland

SpringerBriefs in Computational Intelligence are a series of slim high-quality publications encompassing the entire spectrum of Computational Intelligence. Featuring compact volumes of 50 to 125 pages (approximately 20,000-45,000 words), Briefs are shorter than a conventional book but longer than a journal article. Thus Briefs serve as timely, concise tools for students, researchers, and professionals.

Aleksandr Raikov

Photonic Artificial Intelligence

 Springer

Aleksandr Raikov
Jinan Institute of Supercomputing
Technology
Jinan, Shandong, China

ISSN 2191-530X   ISSN 2191-5318 (electronic)
SpringerBriefs in Applied Sciences and Technology
ISSN 2625-3704   ISSN 2625-3712 (electronic)
SpringerBriefs in Computational Intelligence
ISBN 978-981-97-1290-8   ISBN 978-981-97-1291-5 (eBook)
https://doi.org/10.1007/978-981-97-1291-5

This Springer imprint is published by the registered company Springer Nature Singapore Pte Ltd.
The registered company address is: 152 Beach Road, #21-01/04 Gateway East, Singapore 189721, Singapore

Paper in this product is recyclable.

Dedicated to my parents and teachers.

Acknowledgements

The author is grateful to his wife for her incredible patience. Great thanks to the staff of Jinan Institute of Supercomputing Technology, Shandong Province, China, especially Jingshan Pan and Meng Guo, and my guardian angel, assistant Veina Chen, for their support in writing the book.

I greatly thank Jiang Chen and the company "Openverse Ltd" (http://www.openverse.io) for their persistence in turning some drawings into works of art using the large language model they developed. I greatly thank Arina Pechalnova for the virtuosic and wonderful design of some figures. Thank Liu Luyao for taking part in designing figures.

Special thanks to Dr. Ding Ning and the company "Nantong Triopto PIC Technologies Co. Ltd" (https://www.triopto.com). For their initiatives in constructing the laser setup and deepening my understanding of microlaser peculiarities.

Introduction

In ancient times, the Vedic said, "Words mean nothing". The world-famous Chinese teacher Confucius noted in the "Analects of Confucius": "If language be not in accordance with the truth of a thing, affairs cannot be carried on to success". Modern Indian teacher Shri Shri Ravi Shankar, in his "Celebrating Silence", wrote: "Knowledge is untruth if it is only words".

Artificial Intelligence (AI) models have two semantics: verbal and wordless. The former is created by data, symbols, schemas, logic, images, and words; the latter includes non-verbal free will, inspiration, imagination, emotions, and thoughts. Currently, knowledge is accumulated verbally in the environment of digital computers. In modern AI, words, image processing, deep learning, and generative language models are mainly digitally realised.

However, the possibilities of the digital environment are limited: the number of machine learning parameters is constantly growing and already exceeds trillions, and it is challenging to make digital semiconductor elements more minor than the size of an atom. At the same time, it is becoming increasingly clear that human cognitive processes reflected through AI are as digital (discrete) as analogue (continuous).

Discreteness (digital) is a heritage of the radio lamp era of computer construction. The radio lamp processed the analogue signal unsteadily, but it switched well between 1 and 0. It was convenient to realise these switches with the digital computer while solving various tasks because the theory of discrete mathematics worked quite clearly. The Nyquist–Shannon–Kotelnikov sampling theorem guaranteed the growing accuracy of the discrete calculations with decreasing sampling intervals. However, such an increasing accuracy cannot continue to the ideal level digitally.

AI cannot feel pleasant memories and emotional experiences like humans, transporting them like a "time machine" into the past. Signals in the surrounding reality have a digital and continuous (analogue) form. A person perceives these signals mainly in analogue form. For example, the primary layers of eye receptors work in an analogue way, as well as the endings of nerve fibres that cause muscles to move. Signals only for transmitting data through the body are converted into discrete (pulse) ones. Due to this transformation, the discrete signal speed of movement through the body decreases significantly, and the reliability of transmission increases.

The wordless semantics have an analogue nature and should be processed in an analogue way by a computer without sampling. For this, not all signals coming into the computer from the environment must be reduced to values at the points of reference in time (pulses) and space (pixels) on the curves of continuous signals, as is now done in digital computers. For example, continuous signal processing can be carried out through light, waves, and chemical effects. The depth of feelings, the chaos of thought, cognitive activity, and the transcendental states of the human mind have to be brought from behind the curtain of AI digital restrictions to the analogue way.

With the help of laser beams, it is already possible to perform addition and multiplication operations, which are carried out considering phase characteristics since wave amplitudes are added up, not intensities. Analogue processes of inverting and scaling, Fourier transform, matrix multiplication, and function convolution are performed at light speed.

Advanced (general, strong) AI must help to solve many complex problems. For example, the Big Bang model describes the birth and development of the Universe well, but not from its beginning, and this model is not the only one that describes the Universe's development. The issues with dark matter and dark energy are hanging. Investigations with the help of the James Webb Space Telescope (JWST) give rise to some doubts about the prevailing views on the Universe's history. There are still many mysterious gaps in science, and advanced AI should support their resolution in future. For example, there is no doubt that the second law of thermodynamics on the increase of entropy of a closed system cannot be violated. However, many scientific laws are only sometimes omnipresent—laws of conservation of energy, momentum, and momentum of the amount of motion are the result of various symmetries, and they should be considered absolute; however, the law of conservation of the baryon number operates only within the accuracy of a modern experiment.

It is becoming increasingly apparent that its hitherto imperishable classical digital paradigm should be supplemented with an analogue one on a fundamental level to resolve such scientific problems. AI models that reflect the knowledge of reality are different from reality itself. As philosopher G. Husserl said, there is an abyss of meaning is lying between knowledge and reality. A bridge over this abyss cannot be built rationally. This bridge has only one support—in the present. It is being built on both sides, often in unrelated places, from time to time. A person simultaneously solves direct and inverse problems: from the present to the future and vice versa. Rationality and computer digitalisation help to support this cognitive process.

However, the meaning of things is not entirely formalisable; it is only sometimes friendly with logic and is often the fruit of paradoxes. It originates at the moment of the appearance of two somethings: atoms, things, words, thoughts, events, etc. Only the meeting of two something can generate meaningfulness. To make sense, we act. Actions create changes and changes—differences. With the differences, new meanings are found, which support moving towards the anticipated goal.

The meaning search resembles the effect of optical interference when the diffraction of rays generates a new image reflecting the features of light sources. This book addresses an attempt to advance AI systems considering non-formalisable cognitive semantics using analogue data processing methods. The development of analogue AI systems may rely on achievements in analogue-like disruptive photonic computing, quantum computing, biocomputing, and neuromorphic computing.

This book addresses the *analogue way* of creating photonic AI (PAI), considering that light is most different from the natural structure of the body structure and the human neurosystem. The term "analogue" is used in this book in two senses. Firstly, it means a continuous change of parameters at each point in space. Secondly, it means the object is all points of continuous coordinates, not a discrete (point, digital) representation of information. The modern digital computer paradigm for knowledge management in AI systems must be revised so that the analogue information model considers wordless semantics.

The idea of using light in computing machines appeared simultaneously with the invention of lasers. For several decades, separate nodes of computing systems have been successfully created using optical technologies. The field nature of light determines the numerous advantages of optical technologies for transmitting, recording, processing, and storing information. Photons have a variety of degrees of freedom representing different quantum states such as polarisation, path, time-bin, and frequency. They can utilise high-dimensional or continuous variables. Photons are not subject to decoherence. Photons form electromagnetic waves. They are stable, but the material for image recording is unstable relative to the photon. Many light beams can pass through the same space area, intersect, and not affect each other due to the lack of an electric charge for photons. Using two-dimensional and three-dimensional characteristics of light allows for high density and speed of information transmission. Well-developed technologies exist for generating, manipulating, and measuring photons in space, fibre, and chips.

The book tries to solve the scientific-technical challenge of the continuous representation of reality in AI systems. This makes it possible to practically reset the time of training of AI systems to zero and break the vicious circle, in which time and energy tend to be infinite, to ensure absolute accuracy of calculations in digital computers. The continuous representation of entropy and the signals in the PAI system also allow for overcoming the visible ahead impasse with the performance raising of digital computers.

There are many challenges to creating PAI, including the synthesis of new materials, the control of optical processes, and the building of effective interfaces between discrete and analogue computing.

The book's structure aims to consistently create PAI with enriched cognitive semantics of models described in our previous book [1]. From the beginning of the book, a conceptual formulation of the problem is made. Then, the architecture details of the PAI system's components and a description of photonic materials for rewritable 3D holographic memory are given.

Reference

1. Raikov, A.: Cognitive semantics of artificial intelligence: a new perspective. Comput. Intell. **XVII**, (2021). https://doi.org/10.1007/978-981-33-6750-0

Contents

1	**Impulse Brain?**	1
	1.1 Chapter Conclusion	7
	References	8
2	**Artificial Mind**	11
	2.1 Chapter Conclusion	20
	References	21
3	**Photonic Psychology**	23
	3.1 Chapter Conclusion	30
	References	30
4	**Situational Emotions**	33
	4.1 Strategic Emotions	33
	4.2 Emotions Recognition by Brain Signal or Face	34
	4.3 Emotions Recognition by Text	35
	4.4 Compressing Meeting Time by Considering Emotions	36
	4.5 Implementations	38
	4.6 Chapter Conclusion	39
	References	40
5	**Photonic Thought**	43
	5.1 Chapter Conclusion	49
	References	49
6	**Soliton Thoughts**	51
	6.1 Laser Solitons Recall Human Thoughts	51
	6.2 Laser-Soliton Equipment	54
	6.3 Chapter Conclusion	57
	References	58
7	**Subatomic Thinking**	59
	7.1 Chapter Conclusion	67
	References	67

8   Dark Mind 69
    8.1   Chapter Conclusion 76
    References 77

9   Material Synthesis 79
    9.1   3D Holographic Idea 79
    9.2   AI Modelling 81
    9.3   Graphene Approach 85
    9.4   Quantum Dots 87
    9.5   Inverse-Engineering 88
    9.6   Chapter Conclusion 89
    References 89

10 Photonic Learning 93
    10.1 Digital and Analogue Training 93
    10.2 Double Optical Fourier Convolution 96
    10.3 Experimental Foundations 98
    10.4 Instant Learning 100
    10.5 Chapter Conclusion 103
    References 103

Conclusion 105

Glossary 109

About the Author

Professor Aleksandr Raikov is a doctor of technical sciences, chief scientist of the Jinan Institute of Supercomputing Technology in Shandong Province, China, head of the Artificial Intelligence Department of the National Center of Digital Economy, Lomonosov Moscow State University, state advisor of the Russian Federation of 3rd class, and winner of the Russian Government award in the field of Science and Technology. From 1992 to 1999, he was the chief of the Information and Analytical Department of the Russian Presidential Administration. From 1974 to 1991, he developed information systems for high-level state power. He was a student of academician V. A. Kotelnikov, who was one of the discoveries of the sampling theorem. His research and business interests include artificial intelligence, decision support systems, light transformation, and strategic management. He has published 15 books and about 500 papers in peer-reviewed journals and conference proceedings.

Abbreviations

ACCEL	All-analogue Chip Combining Electronic and Light Computing
AI	Artificial Intelligence
AM	Artificial Mind
CNN	Convolutional Neural Network
CPU	Central Processing Unit
CU	Computing Unit
CUDA	Compute Unified Device Architecture
DFT	Density Functional Theory
EEG	Electroencephalogram
GPT	Generative Pre-trained Transformer
GPU	Graphics Processing Unit
HM	Holographic Memory
HOC	Hybrid Optoelectronic Correlator
JWST	James Webb Space Telescope
LLM	Large Language Model
ML	Machine Learning
PAI	Photonic Artificial Intelligence
PINN	Physics-Informed Neural Networks
QD	Quantum Dot
RNN	Recurrent Neural Network
SC	Situational Centres
USB	Universal Serial Bus
VR	Virtual Reality

Chapter 1
Impulse Brain?

The impulse work of the natural neuron system of the human brain and body provides only part of the cognitive processes. However, modern digital Artificial Intelligence (AI) systems rely on impulse (discrete) technological basis. Thus, a digital computer distorts the analogue (continuous) signals, which are also involved in the cognitive work of the brain. Advanced AI systems must also embrace an analogue ground of cognitive processes more deeply.

Currently, AI systems try to cover an analogue ground of cognitive processes by transforming continuous signals into digital form. Analogue processes can be realised without such transformation using photonic techniques, quantum computing, biocomputing, and neuromorphic computing [1]. This book addresses the photonic one, considering that it is most different from the natural neuron system of the human brain and body. Photons also have essential advantages over other quantum particles:

- different degrees of freedom for encoding quantum states (polarisation, path, and frequency);
- possibility to utilise continuous variables (orbital angular momenta, spatial modes);
- it provides quantum information carriers;
- there are technologies for controlling photons and optics processes in space.

In the photonic way of AI development in the "digital environment", the impulse nature of the human neuron system behaviour must also be considered. An exciting number of scientific and practical works are devoted to the impulsive nature of the brain. It is enough to refer to the book [2] to make sure of this. However, the behaviour of the brain is not limited by impulses. Dreams and intentions, imagination and premonition, mental images of situations, social communications, emotions, attention, fear and its control, a person's idea—it is impossible to describe with only the help of impulses and digits. Less abstract operations, such as pattern recognition and signal transmission, implemented using impulses at first glance, have a more complex character and are not fully reduced to discrete procedures.

The human brain functions with the help of electrical impulses, chemical reactions, waves, and quantum effects. The most common and studied point of view, especially among the creators of AI systems, boils down to the impulse behaviour of brain neurons. The neuron receives several impulses at the entrance and emits an impulse. On average, each natural neuron sends into the axon, departing from it or receiving one impulse per second into each dendrite that enters it [3, 4]. Moving along the axon, the impulse is cloned and comes into contact with thousands of other neurons. Most of the several thousand inputs of a single cortical neuron receive signals from different axons.

So many natural neurons are so diverse in shape and behaviour that building a universal mathematical model is impossible. To do this, artificial neural networks, which, however, very vaguely resemble natural neurons, are constructed and used. It is an impulse technique that AI systems are creating, such as Generative Pre-trained Transformer (GPT), Large Language Model (LLM), genetic or immune algorithms, deep learning, etc. They are expected to imitate the similarity of human thinking.

At the same time, it is known that a natural neuron, unlike an artificial one, fluctuates spontaneously and behaves differently with the same input information. Some neurons, for example, in the first layers of the eye's retina, behave in an analogue way; that is, they do not emit impulses, preferring the transmission of a continuous signal along a chain of neurons. Some neurons behave passively and silently—no situation at the input will force them to generate an impulse at the output. They are like dark energy and dark matter in the Universe—a comparison which may seem absurd, but such a romantic coincidence may be helpful in the research and design of advanced AI systems.

There is a misconception that the impulses of natural neurons are initiated only by external factors. Neurons themselves also have spontaneous activity, which resembles chaotic behaviour, that is, behaviour in which each event has a short memory and weakly affects the next event. There are so-called spontaneous impulses that are generated by natural neurons autonomously, without any signal from the outside, without the influence of other neurons or signals from the world on them. Spontaneous impulses sent by a neuron without incoming impulses create unregulated and unstructured brain activity.

Natural neurons can spontaneously excite and exit impulses due to multiple feedback or cyclic reactions of molecules inside neurons. A single neuron can have about a hundred direct feedbacks through the neurons with which it contacts itself. Moreover, due to this spontaneous behaviour, the output pulse signal of each neuron of a natural neural network may be generated more regularly than the input. In that case, this neural network self-organises and comes to a state where its impulses will be sent periodically [5, 6]. At the same time, irregular, spontaneous impulses make a strategist out of a person, giving him the gift of prophecy and foresight and generating a proactive reaction in a situation of danger. They prevent human thought from falling asleep.

The science and practice of neural imaging, especially with the help of magnetic resonance imaging, are used to study the behaviour of brain regions when processing visual signals, emotional reactions of the face, and activity during sleep. At the same time, each point of the tomography image contains 100,000 neurons. The tomography can register the intensity of the blood flow circulating at this point. However, tomography does not allow us to study the behaviour of individual neurons and the impulses they generate, not to mention the impossibility of studying the behaviour of groups of atoms in a neuron that may be subject to non-local influences.

There is a repeatedly confirmed hypothesis that the activity of a neuron occurs on the principle of "all or nothing"; either it emits an impulse of a given shape, or it does not [7]. An impulse is formed when the potential difference between the neuron body and the environment is sufficient. Ion channels open and close in the membrane, and ions penetrate through them towards each other (sodium ions inside and potassium ions outside); as a result of a sharp increase and then a drop in the resulting voltage, an electric pulse is formed and transmitted to the axon of the neuron. At the same time, the processes in the membrane always proceed the same way. Therefore, the electric pulse has a relatively invariant structure.

The "all or nothing" principle is well described by binary code, which, in turn, implies binary logic, that is, discreteness, which is well supported by a digital computer. This means that a group of neurons, exchanging ones or zeros and emitting pulses depending on different combinations at the input, allows you to build any element and implement logical operations to solve many discrete mathematics problems. To excite a neuron and emit a pulse, it is necessary to receive hundreds of pulses distributed over thousands of input lines to its dendrites. Moreover, the success of the excitation depends on the ratio of the excitation and inhibition pulses, the synchronicity of the input signals, and the location of the dendrites where they fall.

Such a discrete approach gave rise to the trend of von Neumann calculations based on the binary paradigm. This was also facilitated by the architecture of the first computers in which radio tubes (triodes, pentodes) performed discrete operations well, but analogue—bad. It is worth noting that the modern paradigm of creating a quantum computer, based on the use of the Q-bit alphabet, has also not moved so far from the basis of the binary paradigm since quantum parallelism and data are based on binary code, which is only interpolated by the effect of the Q-bit superposition, introducing a corresponding error into the simulated natural analogue signal due to the interpolation.

As noted above, the first layers of visual natural neurons do not use impulses but transmit signals to each other continuously, using voltage control and chemicals—photons of light fly into the cones of the first layer of retinal neurons. When a photon is absorbed, the cone's potential drops and the molecules' flow is suspended. The second layer of neurons captures this moment as a signal and converts it into a change in potential. These first two layers of neurons optically and chemically convert light into voltage without exchanging impulses, so they cannot perform logical operations.

Only the neurons of the third layer of the eye, which are ganglion cells, turn their electrical potential into impulses. The question is, why?

The discrete signal representation impulses allow neurons to transmit information from one place of the human brain or body to another *accurately, quickly, and at the proper distance*. Nature makes signals inaccurate to send them accurately. That is, on the receiving side, neurons process an erroneous signal. To refine it, it is interpolated into a continuous signal to control the work of the muscles. The same approach is observed in computer communication technology for accurate signal transmission, sampling, and encryption. The Nyquist–Shannon–Kotelnikov sampling theorem is used to evaluate and improve the accuracy of continuous signal transmission [8].

The accuracy of the natural neuron work is limited—it takes less than a millisecond to create an impulse. That is, it fixes the time of the event with an accuracy of milliseconds and even microseconds. To evaluate these parameters, experiments were carried out, for example, on rats [9]. An axon departing from a neuron transmits impulses at a high speed, but thousands of times less than the speed of light. In the cerebral cortex, an impulse is sent up to 200 mm per second. In the spinal cord, axons transmit a signal faster—up to 70 m per second [10]. An impulse carries a signal along an axon tens of times faster than by changing the potential and a thousand times faster than through the dynamics of molecules at synapses [11]. At the same time, analogue messages transmitted in the first layers of the eye's retina are sent much slower, and when transmitted over long distances, analogue signals in the human body can be subject to unacceptably significant distortions [12].

At the same time, nature has left the issues of recognition and determining the location of the event to be solved by the analogue mechanism of the eye's retina. It registers light on receptors and turns it into impulses for further transmission to the back part of the brain—the occipital lobe that is involved with vision. The retina records the dynamics of an increase or decrease in the photon flux from the source. The retina also processes images: clustering, matching, and segmenting. Only then is the result of processing through the ganglion cells of the third layer of the retina communicated to the brain. Information about the location of objects perceived by the eye is stored in each layer of retinal neurons, and impulses transmit the perceived image to the brain.

An electric pulse arriving along the axon opens bubbles with molecules so that their contents fall into the gap between the nerve endings of the axon and the dendrites of the receiving neuron through a chemical procedure. The ions will create a voltage surge on the receiving dendrites. The magnitude of the potential jump in the receiving neuron depends on the molecule type. Thus, the pulse again becomes a continuous (analogue) process of releasing molecules, which causes jumps in the cell's potential.

An essential reason for transforming the pulse back into an analogue signal is that transmitting the signal using chemistry and potential is less energy-consuming. The analogue method creates greater flexibility of brain signals. Flexibility is provided by the difference in the structure and behaviour of synaptic slits, which makes it possible to operate with the same discrete code in different ways. The small size of the synaptic system also contributes to the rapid functioning of the analogue data.

Apparently, similar to the analogue scheme of work, not all pulse signals are triggered in the human nervous system; for example, when bubbles with molecules in the synapse end and there is nothing to release during pulse initiation, a voltage surge in the dendrite does not occur. The brain, transmitting information between neurons, allows synapse failures to prevent this transmission. There may be several reasons, such as checking the reliability of communication or monitoring the intensity of the interaction of neurons in "test mode." For example, unreliable synapses help increase information transmission efficiency with minimum energy consumption [13]. Moreover, the more synaptic contacts there are, the more unreliable these synapses work [14].

In this unreliability, nature may have a specific meaning. For example, the unreliability of synaptic contacts creates new ways of signal processing since, for example, a contact failure causes the ability to receive the next pulse to be quickly restored, or a short-term failure serves as a filter for random oscillations at the inputs of dendrites, since failures will ensure the randomness of pulse loss in the flow of wavelike oscillations of signals at the input. In the presence of failures, neurons will no longer be overloaded with fluctuations in input signals, which will help restore control of the behaviour of neurons [15].

Failures in neurons are springs for extraordinary human behaviour, thinking strategically, and solving tricky problems. These failures create the conditions for the success of mental abstraction and generalisation. No matter how "did" they are and how many billions of connections may be added, artificial neural networks still function fairly selectively and stable. In digital artificial neural networks, the more accurate the training, the less complete the coverage of the diversity of the event under study. The artificial network can only adjust its parts to the details of each captured image. A natural neural network has non-comparable greater flexibility and foresight.

An excellent example of the benefits of random choice simulation is a genetic algorithm that randomly builds generations of solutions. Still, it simultaneously steadily converges the process to the goal [16]. Thus, interference helps to solve the inverse problem—a step-by-step process of going to the goal by almost randomly selecting the better solutions.

That is, to solve life problems, nature creates interference in the brain uniquely, as a result of which its adaptability increases, which is followed by the reliability of action in constantly changing external conditions. Synaptic failure in natural neural networks breaks connections randomly. It adds noise to the brain so that the cognitive process does not fall into the trap of overly selective learning, as may be in AI systems.

Different groups of neurons seem to react to something of their own: images, speed, thoughts, etc. For this, they constantly send impulses in response to input signals. But this is not the case. For example, a nerve cell generates five pulses per second; for this purpose, it is initiated arbitrarily by several hundred incoming excitation pulses. In this case, a neuron can have more than 7000 exciting inputs. So, the output pulses should be more than five. However, the cortical neuron can produce no more than 30 pulses per second.

Some neurons mentioned above do not send impulses; they are inactive—"dark" neurons. Studies show that during a specific time interval, the characters of push pulses per neuron resemble a rank distribution. Most neurons will be utterly inactive during this period, but only a few neurons will send the bulk of, say, 90% of the impulses [17]. There is the hypothetical suggestion that such inactive, dark neurons are everywhere in the brain. They take energy for their existence, meaning they are needed for some unknown reason.

One of the ideas about the usefulness of dark neurons is to ensure that they are ready to meet surprises for a sufficiently long period. To do this, the range of possibilities of dark neurons should be broad and, of course, not limited only to the expectation of discrete signals with a limited spectrum of representation. Some of these neurons may not fire or are activated once during a person's lifetime. A possible answer to the need for such neurons may be the ability of a large group of them to generate one pulse at the same time and at the required time, and not a series, as usual.

The romantic idea of understanding the need for dark neurons may be their association with dark energy and dark matter, which comprise about 96% of the Universe. Dark energy, according to the General Theory of Relativity, creates a negative pressure that generates repulsion and antigravity of objects in space, and dark matter explains the observed effects of excessively high rotation speeds of the outer regions of galaxies and gravitational lensing.

Dark neurons can also be called those that send impulses themselves without reacting to anything. Up to 30% of such neurons send impulses unrelated to the solved task [18]. Therefore, the thesis about the possibility of dividing neurons into "responding" or "not responding" after summing up the input effects, taking into account a certain "threshold", is a misconception—the spectrum of diversity of neuronal behaviour is much broader than binary.

However, the impulsive behaviour of natural neurons is most often used to create AI systems, where impulses are sent to the input of an artificial neuron. As a result of their processing, this neuron can be excited and generate an impulse, which is transmitted to the following neurons. The transformation of natural continuous signals in the human neuron system and distortion of continuous signals in the digital AI system is illustrated in Fig. 1.1 (designed using the LLM by Openverse Ltd).

Figure 1.1 also shows the human body's combination of pulse and analogue waveforms. The continuous signal from the external environment is transmitted to human receptors in analogue form, pre-processed (clustering, classification, recognition) and then converted into a pulse signal for transmission over a relatively longer distance through the human brain and body. In the human brain, the signal is processed in pulse and analogue form; the effect on the muscles is carried out in the analogue format.

Discretisation of the signal in natural neural neurons introduces errors into it and distorts its spectrum. Such errors are insignificant for controlling muscles in a natural environment. Although nature introduces randomness and errors into the behaviour of human neural systems, it makes a person more creative and strategic.

However, in AI digital modelling of cognitive processes, when it is necessary to consider the non-formalised mental, emotional, and spiritual aspects of consciousness, introducing digital errors and distortion of the signal spectrum reduce the capabilities of AI systems to represent natural human cognitive functions.

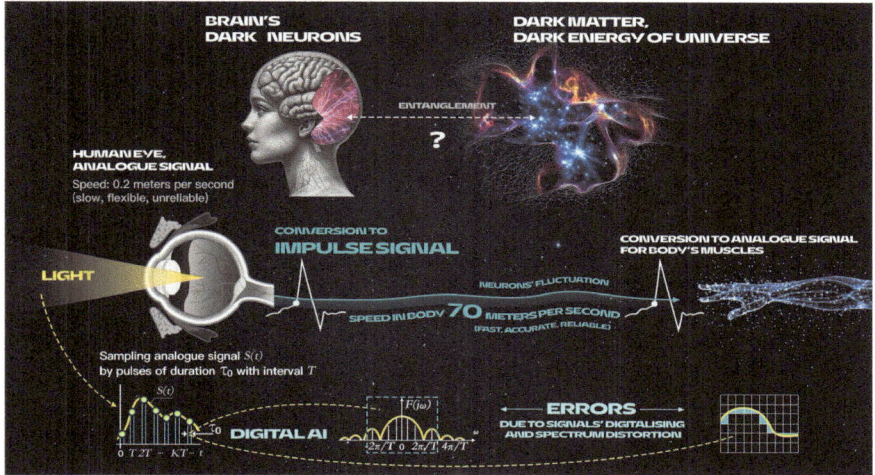

Fig. 1.1 Transforming and distortion of a continuous signal in natural systems and digital AI (published with permission by Openverse Ltd)

Figure 1.1 illustrates the fractal analogy of the Universe's and the human brain's shadow ("dark") sides. Such analogies can help to answer complex questions about the human mind's nature in the context of the Universe's birth and behaviour.

1.1 Chapter Conclusion

- A natural neuron (the retina of the eye, synapses, nerve endings on the muscles of the body, etc.) performs some functions through chemical processes in an analogue (continuous) way; an artificial one—processes data digitally (discretely).
- In a natural neuron, an analogue signal propagates slowly (0.2 m/s), flexibly and unreliably; in an artificial one—data is processed at the speed of an electromagnetic wave in a cable (about 2/3 of the speed of light), discretely and reliably.
- Signals are transmitted by pulses at a speed of up to 70 m/s between natural neurons quite accurately and reliably; between artificial neurons, signals are transmitted in binary form at the rate of an electromagnetic wave in a cable.
- The human brain contains many "dark" neurons; the purpose and functions are still at the level of hypotheses, which resembles the situation of dark energy and dark matter in the Universe; an artificial neural network has a transparent structure with determined behaviour.

- To transmit data by pulse, an error is introduced into a continuous signal, and, accordingly, its spectrum is distorted due to the difference in the values of analogue and discrete signals since values to the latter are set only at the sampling points (the Nyquist–Shannon–Kotelnikov theorem).
- The reason for transforming the pulse signal back into an analogue one is that the latter is less energy-consuming and creates greater flexibility and richer semantic interpretability of brain signals.
- The advanced (general, strong) AI must consider the analogue nature of natural neurons' behaviour, including atoms in a neuron and atomic spontaneous, non-local, and fluctuation effects.
- The perception and first processing of continuous signals (light, photons) are realised in an analogue way without discrete sampling and simultaneously consider the atomic effects of the behaviour of natural neurons.

References

1. Zhu, S., Yu, T., Xu, T., et al.: Intelligent computing: the latest advances, challenges, and future. Intell. Comput. **3**, 0006 (2023). https://doi.org/10.34133/icomputing.0006
2. Humphries, M.: The spike: an epic journey through the brain in 2.1 seconds. Princeton University Press, Princeton (2021)
3. Laughlin, S.B., Sejnowski, T.J.: Communication in neuronal networks. Science **301**(5641), 1870–1874 (2003). https://doi.org/10.1126/science.1089662
4. Lennie, P.: The cost of cortical computation. Curr. Biol. **13**(6), 493–497 (2003). https://doi.org/10.1016/s0960-9822(03)00135-0
5. van Vreeswijk, C., Sompolinsky, H.: Chaos in neuronal networks with balanced excitatory and inhibitory activity. Science **274**, 1724–1726 (1996)
6. van Vreeswijk, C., Sompolinsky, H.: Chaotic balanced state in a model of cortical circuits. Neural Comput. **10**, 1321–1371 (1998)
7. Arbib, M.A.: Warren McCulloch's search for the logic of the nervous system. Perspect. Biol. Med. **43**(2), 193–216 (2000)
8. Raikov, A.: Cognitive semantics of artificial intelligence: a new perspective. Topics: Computational Intelligence XVII. Springer, Singapore (2021). https://doi.org/10.1007/978-981-33-6750-0
9. Bale, M.R., Campagner, D., Erskine, A., et al.: Microsecond-scale timing precision in rodent trigeminal primary afferents. J. Neurosci. **35**, 5935–5940 (2015)
10. More, H.L., Hutchinson, J.R., Collins, D.F., et al.: Scaling of sensorimotor control in terrestrial mammals. Proc. R. Soc. B Biol. Sci. **277**, 3563–3568 (2010)
11. Sterling, P., Laughlin, S.: Principles of neural design. MIT Press, Cambridge (2015)
12. Romand, S., Wang, Y., Toledo-Rodriguez, M., et al.: Morphological development of thick-tufted layer V pyramidal cells in the rat somatosensory cortex. Front. Neuroanat. **5**, 5 (2011). https://doi.org/10.3389/fnana.2011.00005
13. Levy, W.B., Baxter, R.A.: Energy-efficient neuronal computation via quantal synaptic failures. J. Neurosci. **22**(11), 4746–4755 (2002). https://doi.org/10.1523/jneurosci.22-11-04746.2002
14. Branco, T., Staras, K., Darcy, K., et al.: Local dendritic activity sets release probability at hippocampal synapses. Neuron **59**(3), 475–485 (2008). https://doi.org/10.1016/j.neuron.2008.07.006

15. Rosenbaum, R., Zimnik, A., Zheng, F., et al.: Axonal and synaptic failure suppress the transfer of firing rate oscillations, synchrony and information during high frequency deep brain stimulation. Neurobiol. Dis. **62**, 86–99 (2014). https://doi.org/10.1016/j.nbd.2013.09.006

16. Raikov, A.: Convergent fuzzy cognitive modelling of regional youth policy strategy. In: Yang, X. S., Sherratt, R. S., Dey, N., Joshi, A. (eds) Proceedings of Eighth International Congress on Information and Communication Technology. ICICT 2023. Lecture Notes in Networks and Systems, vol. 694, pp. 911–921. Springer, Singapore (2023). https://doi.org/10.1007/978-981-99-3091-3_74

17. Wohrer, A., Humphries, M.D., Machens, C.: Population-wide distributions of neural activity during perceptual decisionmaking. Prog. Neurobiol. **103**, 156–193 (2013). https://doi.org/10.1016/j.pneurobio.2012.09.004

18. Maggi, S., Peyrache, A., Humphries, M.D.: An ensemble code in medial prefrontal cortex links prior events to outcomes during learning. Nat. Commun. **9**, 2204 (2018). https://doi.org/10.1038/s41467-018-04638-2

Chapter 2
Artificial Mind

The idea of the Artificial Mind (AM) can denote a new step in developing the Artificial Intelligence (AI) paradigm. Currently, the AI paradigm relies on digital approaches. However, progress in AI development is primarily determined by interpreting the mind's analogue—non-digital and non-local—thinking processes.

So far, the main development of AI has been based on digital computers. This AI allows for good autoregression, pattern recognition, classification, text analysis and synthesis, etc. The improvement of AI is mainly carried out due to the extensive growth of computing power, volumes of data sets, and the accumulation of different heuristics. This is not so little; it even looks like a scientific and technological revolution, but it is just accelerated evolutionary progress within the classical digital paradigm, which was suggested last century.

However, the crucial limitation of the computer digital paradigm is well known. The relation of the digital and continuous computer paradigms to nature is detailed in [1]. In particular, the fallacy of giving a digital interpretation to continuous entropy is shown. The continuous entropy does not need bits to encode a natural system, although it quantifies its information content. The natural system can be encoded with a finite number of bits, which is only inaccurate.

The dynamics of any artificial system's development are often influenced by its name. For example, the term "intelligence" gives the phenomena of the human mind, thinking, and consciousness a shade of rationality, logic, optimisation, and formalizability. That is, the very name of AI drives its development into formalised blinders. On the other hand, mind, thinking, and consciousness have more crucial informal content. Therefore, the depth of feelings, the chaos of thought, and the cognitive and transcendental inherent in the phenomena of mind must be brought out from behind the curtain of the formalised (digital) restrictions. In the following exposition, the concepts of mind, thinking, and consciousness, considering that we are talking about an unformalised fragment of these concepts, will provisionally and conditionally be regarded as synonyms.

© The Author(s), under exclusive license to Springer Nature Singapore Pte Ltd. 2024
A. Raikov, *Photonic Artificial Intelligence*,
SpringerBriefs in Computational Intelligence,
https://doi.org/10.1007/978-981-97-1291-5_2

The phenomenon of thinking is the activity combined with human capability, characterised by a generalised and indirect reflection of reality. Its basis is the emergence and replenishment of various concepts, ideas, conclusions, and judgments. In the classical digital paradigm of AI, thinking, as a process, seems formalised and is usually associated with a computer, an intelligent machine. However, a person has a much greater potential for thinking fuelled by feelings and movement activity through the physical and biological interaction with the environment, the cosmos, due to the non-local characteristics of neurons' subatomic behaviour.

The essence of thinking is in performing cognitive operations with symbols, texts, images, feelings, and transcendental states of a human's mind. This makes it possible to build verbal models of the world, thanks to which the representation of the world becomes more perfect. The words join the image of the subject of thinking, highlight its essential features that are directly inaccessible to humans and translate the image's subjective meaning into a system of meanings, making it more understandable.

Rituals and symbols and their meanings mean a lot in the picture of the world. For example, in China, a triad of a vase, a crane, and a deer (六合太平) in front of a palace with a legendary and centuries-old history means well-being and harmony (Fig. 2.1).

Hieroglyphs and letters are semantically different. Hieroglyphs are symbols that partially convey the meaning of words through images, sometimes in the form of images, occasionally hinting at their sound but always bypassing their exact reading in sounds. Letters act as abstract signs of sounds. They make up the alphabet, the base of the sound writing code. There is no alphabet for speech sounds in the hieroglyphic language, and the base of the hieroglyphic writing code in everyday practice ranges from two to five thousand units of meanings of monosyllabic words. The base of the hieroglyphic code is about 40,000 characters—considering ancient texts.

Fixation of problems in situations of perception or generation of argumentation in the process of thinking is necessary for forming various personal constructs. The cognitive style denotes the availability of observation features for obtaining, processing and semantically interpreting information about natural cognitive structures. It is essential to distinguish between meaningful and structural aspects of the cognitive sphere in the thinking processes. The significant variables are representations, attitudes, and values. The structured variables are rules for organising, selecting, combining, and linking events.

The study of the nature of thinking usually proceeds from the distinction between sensory and rational cognition, the difference between thinking and perception. The latter reflects the surrounding world in images and provides them with sensually authentic properties. Thinking begins where sensory cognition finishes. Thinking develops the cognitive work of what sensations, perceptions, and representations have given. At the same time, sensory cognition and thinking mutually influence each other and place one into the other.

Interest in the processes of thinking has remained strong since ancient times. For example, considering the essence of the ancient method of immersing a person in a meditative state highlights the following layers of this immersion. First, a person is aware of the external environment the person perceives, separating himself from

Fig. 2.1 Harmony (the author made the photo)

any external phenomena: wind noise and conversations, the smell of flowers, and the behaviour of other external objects. The next layer of immersion is words, texts, reasoning, analysis of phenomena, and planning—the attributes of a person's inner sphere; without violence to oneself, a person also slowly abstracts. The next layer is thoughts; they come and go away. It becomes even more vital if a person tries to weaken some thoughts; a person must let them go, and they will go away. The next layer is feelings, emotions, and sensations; a person must also abstract from them and let them exist alone. The deepest layer is meditative—it may come after uttering a personal mantra. A person falls into it like an abyss of pleasure, and this state cannot be described in words. Meditation is bringing a human into a state between sleep, dream, and wake. In this immersion, only one layer is represented verbally (by words or discrete). However, all layers affect the thinking process.

Thinking is object-oriented. It usually has a focus and a goal: to buy a product, prove a theorem, explain a phenomenon, etc. External sources of information, reasoning, floating feelings, and a colleague hint may be used in the thinking process. Can this be mathematically modelled? There are works on modelling the unconscious. Such works are usually psychological, far from physical and mathematical disciplines, involving cognitive psychology or neurolinguistic programming methods. It is also worth noting that there are studies in quantum psychology in which analogues of various still mysterious natural phenomena for classical physics are considered [2, 3].

For example, the paper [4] concerns the tensor product structure of the high-dimensional Hilbert space, entanglements, superposition, and decoherence for modelling the behaviour of the cognitive state of mind, from which different predictions can be made. Most such works deal with choosing the decision from many alternatives with known likelihoods. However, in real life, the alternatives' probabilities are unknown, or the decision isn't the choice from previously known different options. The decision can be non-alternative—for example, it may be an inverse problem-solving process initiated by an inaccurate target.

It may be necessary to turn to quantum-wave and relativistic theoretical and practical phenomena accompanying the behaviour of living and inanimate, considering the subatomic level of the structure of the human brain's cells and molecules. In addition to the logical and neural network components, AI modelling of the thinking process may accompany various physical patterns that affect the thinking process with varying strengths. For example, the force of gravitational attraction of two protons is about 10^{37} times weaker than the force of their electrostatic repulsion. However, excluding such relatively "weak" forces of influence on individual atomic elements is unnecessary since the number of such elements and physical connections between components are enormous, considering people's motor and mental activity.

In thermodynamics terms, the object-oriental thinking process can be represented by the ability to limit the number of degrees of freedom and preserve the complex behaviour of the thinking process, combining order and chaos [5]. However, a divergent thinking process, especially brainstorming, makes the number of degrees of freedom infinity, thereby creating a continuum that is no longer characterised discretely but continuously. Accordingly, at least a discrete index i and a continuous variable x represent it. To begin, a description of such a system in the form of some potential function $u(x,t)$ can be done in partial differential equations with two independent variables, space point x and time t. In the case of a continuum, we should expect very complex behaviour with the manifestation of chaos, uncertainty, and nonlinearity. However, object-oriented thinking would end non-resultantly if the chaos of thought were allowed to "roam around" indefinitely. Hence, we must find conditions that lead the specified complexity to a subject-determined purposeful regularity.

The regularities that can take place in the AI modelling of the thought process include such phenomena as the behaviour of a laser soliton, that is, a stable solitary wave, or the distribution of energy in a chain of nonlinear oscillators when with an infinite increase in the number of their energy is evenly distributed between modes (see also Chap. 6 and [6]). Nonlinear models show that laser soliton's solitary waves with a more significant amplitude move faster than waves with a smaller amplitude and, at the same time, do not lose their shape; the waves have a slope, the magnitude of which is proportional to the value of the initial conditions, eventually leading to a shock wave. The solitary waves themselves can be associated with an influx of thought, which, the more you think about something, the clearer and stronger it becomes.

A guess, an instant insight, or an idea are non-monotonic phenomena; they may be more associated with a shock wave—a kind of rupture surface of a moving wave medium. In a shock wave, the characteristics of the wave experience a jump, which, unlike the soliton, dissipates quite quickly—the idea came and is already being worked out for practical implementation. Such a regularity model can be a one-dimensional nonlinear chain formed by equal masses. Neighbours are connected by a simple force dependence, for example, of the second order (characteristic of diffusion equations) on the distance between neighbours. Despite their complexity, such model examples are only a simplified slice in the infinite-dimensional space of the phenomenon of thinking. At the same time, studying such models generates new ideas for the further development of AI. For example, the laser wave-soliton idea may help to create the AI models' weakly formalised cognitive semantics [5].

Considering, on the one hand, the infinite number of freedoms in the behaviour of a "thinking" chaotic dynamic system and, on the other hand, the need to limit them to obtain an objective result of thinking, it is advisable to identify the possibility of constructing certain integrals of motion, or conservation laws that characterise the stability and objective purposefulness of the dynamics of a chaotic thought process. Such a law, for example, may reflect the dynamics of the density D_t of some chaotic flow F_x. It helps to find the solution function—the potential function $u(x,t)$—of those mentioned above partial differential equation, which describes the AI system behaviour represented by a discrete index and a continuous variable in the form [7]:

$$D_t + F_x = 0. \tag{2.1}$$

Suppose the density D_t is the gradient of the flow F_x. In that case, the relation (2.1) is valid since both terms are mutually destroyed, which can be considered the first conservation law—in a closed system, the sum of the kinetic and potential energy of the bodies and the forces of gravity and elasticity interacting with each other remains unchanged.

The second conservation law may correspond to momentum conservation. To date, about ten such conservation laws have been identified for chaotic media with waves, but an increase in the number of degrees of freedom has yet to be noticeable. However, it is evident due to common sense that since the high number of degrees of freedom prevents getting the purposefulness (converging to goals) of the thinking process, this number should be limited, and the number of conservation laws also must be limited, for example, by an "observable" amount for an ordinary person.

To give mathematical rigour to the common-sense purposeful behaviour, one can turn to methods for inverse problem-solving on topological spaces, which can be used to find the necessary conditions for converging collective creativity processes to goals [8]. One of these conditions is to ensure the compactness of the space decomposition of the available set of material and intellectual resources to achieve the goal, that is, the finite coverage of this set. Intuitively, this finiteness is limited to a small (observable) number of thematic subsets. However, the interpreting set of resources must be infinite (that is, an infinite number of elements must represent it). Otherwise, another condition for ensuring the convergence of the thinking process to the goal will be violated: its Hausdorffness separability or the possibility of a separable representation of the resources set's elements.

The existence of conservation laws, that is, integrals of motion, in the case of conservation of density, means the existence of some symmetry or invariance. Then, the evolution of the potential function $u(x,t)$ in time can be studied by solving a quantum mechanical problem with the Schrodinger equation. To do this, it is necessary to solve the *direct scattering problem* under initial conditions $u = u(x,0)$, where x is a particular variable, with the search for eigenvalues and eigenfunctions. As *one* evolves as a function of the variable t, the quantum mechanical characteristics of the process, sometimes called *scattering data*, also change.

In this case, the variable t does not carry the time character—it is a deformation parameter. Suppose $u(x,0)$ is the initial potential of the evolutionary process. In that case, the scattering data corresponding to a particular value of t can be used to determine the potential function $u(x,t)$, which implies solving the *inverse scattering problem*.

The proposed physic-mathematical interpretations of possible thought processes need further practical implementation on some physical devices. An optical (photonic) system can be such a device, assuming the fastest and continuous form of data transformation. Creating such a system may be based on an analogue approach with Fourier transforms, which may resemble the solution of linear evolutionary equations. This approach can be introduced in a simplified form as follows (see also [7, 8]).

In the interval $-\infty \leq x \leq \infty$, a transform equation for a function of variables x and t, characterising a certain "potential" of the thought process can have the following form:

$$u_t = \Im\left(\frac{\mathrm{d}}{\mathrm{d}x}\right)u, \tag{2.2}$$

where $\Im\left(\frac{\mathrm{d}}{\mathrm{d}x}\right)$ is a polynomial in $\frac{\mathrm{d}}{\mathrm{d}x}$, that is, a linear operator. For example, with a second-order derivative, Eq. (2.2) is a diffusion equation, which may have a natural association with the interpenetration of various elements of the thought process. Then, the direct Fourier transform for the "potential" $u(x,t)$ is defined as:

$$\tilde{u}(k, t) = \int\limits_{-\infty}^{\infty} u(x, t)\,\mathrm{e}^{-ikt}\mathrm{d}x. \tag{2.3}$$

The inverse Fourier transform will have the following form:

$$u(x,t) = \frac{1}{2\pi} \int\limits_{-\infty}^{\infty} \widetilde{u}(k,t)\, e^{ikx}\, dk. \tag{2.4}$$

The Fourier transform of Eq. (2.2) gives an evolutionary equation for $\widetilde{u}(k,t)$:

$$\frac{d\widetilde{u}}{dx} = \Im(ik)\,\widetilde{u}, \tag{2.5}$$

which has a solution specifying the evolution of the Fourier data for any parameter value t:

$$\widetilde{u}(k,t) = \widetilde{u}(k,0)e^{\Im(ik)t}. \tag{2.6}$$

The initial data $\widetilde{u}(k,0)$ are determined by the given initial conditions $u(x,0)$ using the transformation (2.2). The value of $u(x,t)$ can be obtained from (2.5) using the inverse Fourier transform according to the formula (2.3). As a result, a simplified illustration of the interpretation of the evolution of the thought process considers the effects of randomness and diffusion.

Considering the requirement mentioned above to consider quantum mechanical effects, the Fourier analogy shown in real life will be much more complex since the evolutionary equations become nonlinear. Solving direct and inverse problems when such nonlinear effects occur is necessary. For example, the one-dimensional Schrodinger equation for eigenfunction ψ_x, a potential $u(x,t)$ with an eigenvalue λ has the form:

$$\frac{d^2\psi_x}{d^2 x} - (u(x,t) - \lambda)\,\psi_x = 0. \tag{2.7}$$

This problem can admit coupled quantum states depending on the type of initial value $u_0(x) = u(x,0)$. Equation (2.7) defines a finite set of discrete eigenvalues $\lambda_n = -k_n^2$ $(n = 1, \ldots, N)$, where k is the wave number (the spatial frequency of the wave measured in cycles or radians per unit distance) corresponding to the coupled states (called the spectrum of coupled states), and their corresponding eigenfunctions are ψ_n. For quadratically integrable eigenfunctions of such states, the normalisation condition must be met:

$$\int\limits_{-\infty}^{\infty} |c_n \psi_n(x)|^2 = 1, \tag{2.8}$$

where c_n—is the normalisation constant. For positive $u_0(x)$, the Schrodinger equation defines a continuous spectrum of coupled states with $\lambda = k^2$.

The natural thought process usually repeats many times and in different variants, which can be interpreted as a multiple collision of a quantum wave with potential barriers, reflection from these barriers and subsequent superposition of waves. Then, the interpretation of the thought process through quantum-wave functions will require considering the effect of reflection from these potential barriers, which, in turn, generates the result of the superposition of forward and reverse quantum waves. For one of the potential barriers, respectively, considering the incident e^{-ikx} and reflected e^{ikx} waves, as well as the *reflection coefficient* from the barrier $b(k)$, the expression for $\psi(x)$, will have the form:

$$\lim_{x \to \infty} \psi(x) = a(k)\,e^{-ikx} + b(k)\,e^{ikx}. \tag{2.9}$$

For $x \to -\infty$, the second term in (2.9) tends to 0, and only the past wave e^{-ikx} remains proportional to a specific *coefficient of passage* $a(k)$. This wave has a spectrum and interferes with other waves that have passed through other barriers. Going around and passing inhomogeneous barriers, the wave undergoes diffraction.

Thus, the thought process at the atomic level is characterised by the superposition (with the effects of interference, diffraction, absorption, addition, distortion, aberration, etc.) of a huge, rather even infinite, number of quantum waves, laser solitons, and other wave phenomena. An approximate estimate of the minimum number of waves—if each wave is associated with only one atom—even in one person's brain is 13–15 orders (the number of atoms in one neuron) of magnitude higher than the number of neurons in the brain. The interaction of such waves generates, respectively, an exponentially growing complication.

Thus, the *scattering data* corresponding to the human mental potential $u(x,t)$ represents the sets of all eigenvalues of the coupled states λ_n, the normalisation constants c_n, as well as the continuous functions $a(k)$, $b(k)$. These data can be used to define the function $u_0(x)$ uniquely [9, 10]. Assuming that thinking is substantive, that is, space and, accordingly, the primary energy of a complex wave process, is limited to a specific volume, $u_0(x)$ must satisfy the condition:

$$\int_{-\infty}^{\infty} (1 + |x|)|u_0(x)|dx < \infty. \tag{2.10}$$

For a passing wave, considering the relation (2.10), it is possible to determine a value that can be regarded as some Fourier transform of the scattering data:

$$B(\zeta) = \sum_{n=1}^{N} c_n^2 e^{-ik\zeta} + \frac{1}{2\pi} \int_{-\infty}^{\infty} b(k)\,e^{ik\zeta}\,dk \tag{2.11}$$

Further, the function $K(x,y)$ is introduced, which is found through the linear integral equation:

$$K(x, y) + B(x, y) + \int_{-\infty}^{\infty} B(x + z)K(x, y)\,dz = 0, \tag{2.12}$$

and,

$$u_0(x) = -2\frac{\mathrm{d}}{\mathrm{d}x}K(x, x). \tag{2.13}$$

That is, the potential $u_0(x)$, which sets the scattering data included in the formula (2.12), with its "deformation" to $u(x,t)$, can be calculated by the formula (2.13). Equation (2.12), in relation to the actual situation with the interpretation of the thinking process, can be solved strictly for a very degenerate case.

Thus, these transformations can be supported using optical devices that provide the construction and convolution of Fourier holograms. Holographic devices allow for compact and reliable storage, recording and reading of information on a holographic storage device (see Chap. 10). In this case, compactness may also imply multiple recording and playback of various images at one point of the drive due to implementing the Fourier convolution operation of a set of images. To obtain such holograms, a flat modulating half-transparent screen with the x and y axes, carrying information about the modulating object, represented by the function $f(x, y)$, is lighted by the expander of the laser beam. The modulated beam (object wave) is directed at the lens and focused from its other side. A holographic storage device is placed in the focal plane of the lens to form a point (spot, dot) with a Fourier image of the recorded object. A reference beam is directed to the same point of the storage at a certain angle θ, which interferes with the object wave, forming a corresponding Fourier hologram at this point.

The lens provides a frequency analysis of the function $f(x, y)$, which represents the recorded object, replacing it with harmonics of spatial frequencies. Each point of the recorded hologram is characterised by its spatial frequency and phase. Since the reference wave falls on the recording plane at a certain angle, thereby fixing the spatial frequency spectrum for a particular recorded object, changing the angle of incidence changes the interference pattern. This effect can be used to write another object to the same point (is also known as angle multiplexing or angular multiplexing). However, in this case, the problem moves to the field of synthesis of a unique photonic material that will allow the storage of many different interference patterns at one point (see Chap. 9).

Optical Fourier transforms have many valuable properties. Thus, the compactness of information storage when using optical devices and Fourier transforms is ensured by compressing images in the focal plane of the lens over a small area and, if appropriate material is available, repeatedly recording and storing multiple images at one point. Storage reliability is ensured by the possibility of saving information in case of loss of a part of the spectrum or violation of a part of the carrier since information about each point of the recorded image is contained at any point of its Fourier image. With such distortions, the resolution and brightness of the image are not significantly reduced. Such holograms have less aberration compared to other types of holograms.

Fourier holograms can serve as modulating filters in Photonic AI (PAI) systems, significantly speeding up the learning process of an analogue optical neural network compared to its digital version. Holographic recording has associative properties. That is, it allows to restore the complete image of an object, for example, the text of a page, according to its part. In this case, the intensity of illumination of the recorded image in the output plane is invariant to the displacement of this image on the modulating half-transparent screen in some direction.

A special place in the creation of PAI can be occupied by the convolution of functions reflecting the images on the modulating half-transparent screen. Such an operation can be interpreted as a comparison of functions. The convolution operation can be applied to a set of Fourier images, which is necessary for training an optical neural network (see Chap. 10). This property of convolution can be used in the construction of an optical neural network, in which the comparison of two parameters set in the form of images plays a crucial role. For two functions, for example, $f(x)$ and $s(x)$, convolution can be performed through multiplication and Fourier transform operations. To achieve the convolution operation $g(x) = f(x) \otimes s(x)$, it is necessary to perform the Fourier transform on the functions $f(x)$ and $s(x)$, multiply the resulting images and perform the inverse Fourier transform on the resulting product. In such a way, the convolution of a linear combination of a finite number of images can be realised using the addition, multiplication, and Fourier transform operations.

These calculations and optical representations, taking into account atomic structure and effects of human neuron behaviour, illustrate the processes of thinking that lie much deeper than the logical, digital, and modern artificial neural network interpretation of the behaviour of the human brain and body. Such interpretive processes can be used to illustrate and, possibly, hardware implementation of weakly formalised cognitive semantics of the next generations of AI (strong, general) [5]. These generations must be much semantically richer than traditional machine-like understanding of digital AI phenomena, and, as an idea, the term "intelligence" may be changed to "mind".

In this context, a change of digital paradigm is needed for the further fundamental development of AI. Considering the informal (continuous) aspects of mind, thinking, and consciousness with their implicit, non-causal and non-local effects is necessary. Artificial Mind may denote this step of AI development.

2.1 Chapter Conclusion

- The development of traditional AI is based on the formalisable digital version of computing and retrospective data, allowing autoregression, pattern recognition, classification, analysis, and synthesis of texts.
- Currently, the essence of thinking is in performing cognitive operations with symbols, texts, voice, and images, making it possible to build formal models however this is a limited part of the thinking phenomenon because it doesn't consider deep feelings and transcendental states of mind.
- It is necessary to consider the informal aspects of the human mind with its implicit, non-causal and non-local effects—for the further fundamental development of AI, its general, strong, and perhaps Artificial Mind version.

- The ancient method of immersing a person in a meditative state of mind highlights some layers of the immersion; however, currently, digital AI covers only part of these states of mind.
- A guess, an insight, or an idea are non-monotonic phenomena associated with a shock wave, a kind of rupture surface of a moving wave medium.
- Thermodynamic, quantum, and wave approaches illustrate the thinking processes that lie much "deeper" than the logical and neural network interpretation of the behaviour of the human brain by traditional AI.
- An advanced version of AI (general, strong) must be much semantically richer than digital AI, so the term "intelligence" may be changed to "mind" for this identification.

References

1. Lee, E.A.: Plato and the Nerd. The Creative Partnership of Humans and Technology, p. 288. The MIT Press, Cambridge (2018)
2. Haven, E., Khrennikov, A.: Statistical and subjective interpretations of probability in quantum-like models of cognition and decision making. J. Math. Psychol. **74**, 82–91 (2016). https://doi.org/10.1016/j.jmp.2016.02.005
3. Pothos, E.M., Busemeyer, J.R.: Quantum cognition. Annual Reviews of. Psychology **73**, 749–778 (2022). https://doi.org/10.1146/annurev-psych-033020-123501
4. Dorje, C.: Brody. Quantum formalism for cognitive psychology (2023). https://doi.org/10.48550/arXiv.2303.06055
5. Raikov, A.: Cognitive semantics of artificial intelligence: a new perspective. In: Topics: Computational Intelligence XVII. Springer, Singapore (2021). https://doi.org/10.1007/978-981-33-6750-0
6. Grelu, P., Akhmediev, N.: Group interactions of dissipative solitons in a laser cavity: the case of 2+ 1. Opt. Express **12**(14), 3184–3189 (2004). https://doi.org/10.1364/opex.12.003184
7. Tabor, M.: Chaos and integrability in nonlinear dynamics: an introduction. John Wiley & Sons, Inc., New York (1989). https://doi.org/10.5860/choice.27-2142
8. Raikov, A.: Convergent ontologization of collective scientific discoveries. In: Proceedings of the 14th International Conference Management of large-scale System Development (2021). https://doi.org/10.1109/MLSD52249.2021.9600184
9. Marchenko, V.A., Agranovich, Z.S.: Potential recovery from the scattering matrix for a system of differential equations. Dokl. USSR Acad. Sci. **113**(5), pp. 951–954 (1957) (in Russian)
10. Gelfand, I.M., Levitan, B.M.: On the definition of a differential equation by its spectral function. Izv. USSR Acad. Sci. Ser. Math. **15**(4), 309–360 (1951) (in Russian)

Chapter 3
Photonic Psychology

Psychological practices can be accelerated, and their effectiveness can be increased with the help of advanced Artificial Intelligence (AI) systems. Psychology is a subtle science that delves into the nuances of a person's mental life and emotions, which digital AI cannot fully embrace; however, photonic AI can due to the possibility of considering such nuances of the human's signals.

Weak AI is relatively rude; it works with digital information and limited signal spectra. Optical (photonic) methods in recognising people's behaviour, gait, emotions, etc., can form a booming field of application AI for psychological science and practice. For example, let's consider applying AI methods to the organisation of a six-step reframing procedure—from the psychological field of neuro-linguistic programming. We have used this procedure to organise the team's strategic planning meeting [1]. Currently, the strategic meeting can consist of some sessions with more than 5 h each, but it is too long to keep participants on a high level of creativity and cannot be applied in an emergency [2].

Figure 3.1 shows the scheme of the psychological six-step reframing. Possible emotional expressions of faces and designations of texts (publications) that participants previously wrote have been added to the figure. AI tools can use these elements to assess participants' behaviour and emotional mood during a strategic team meeting. In strategic reframing, the moderator of the meeting closely monitors the members, their behaviour, the manifestation of emotions, etc. If the reframing is carried out in a group, it is almost impossible to follow the participants' behaviour deeply enough. Computer vision and machine learning systems can help. However, modern computer systems work in digital format, and only about ten types of emotions are recognised on a discrete scale. Still, the moderator needs to get more information about participants' feelings. It is necessary to monitor the time dynamics of participants' emotions changing on a continuous emotional scale with the diagnosis of various nuances of emotions coupled with events.

Currently, the quality of discrete emotion recognition by digital computer vision and AI reaches 90–91%. Since emotion recognition with AI support is carried out on

1. Identify collective subject X moving toward the goal.

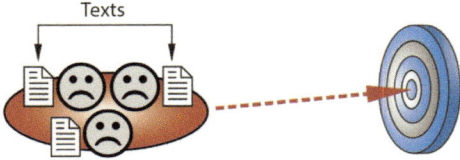

2. Establish communication with the part responsible for X (leader) and recognise his/her intention.

3. Separate the behaviour of X from the positive intention of the part responsible for X.
4. Ask X to generate 3 new ideas for achieving the goal.
5. Ask the part of X if they want to take responsibility for using new ideas.
6. Environmental audit.

Fig. 3.1 Scheme of six-step reframing

a discrete (digital) scale, this fundamentally distinguishes it from the natural mani-festation of people's emotional behaviour. Emotions influence human continuous (analogue) analytical thinking. However, the use of discrete data processing mech-anisms in AI involves dividing the image or problem into parts, small and large, which distorts the meaning of source data, accumulates errors during processing, juxtaposes details, and, as a consequence, the loss of the essence of the image or problem can emerge.

In an actual cognitive situation, spirituality, confusion, reasonableness, and other qualities of a person impose on the emotional and analytical processes a sense of integrity a priori perception of the whole. With a mechanical digital analytical approach, it can become an unsolvable mystery which motive had the main impact on intuitive decision-making by a person—because in the process of digital analysis, this motive and its connections can be split and lost.

Creating digital AI systems for emotion recognition hasn't been focused on how human beings holistically think and feel. AI researchers have focused on formal-isable factors, including digital vision, algorithms, models, data analyses, logic, samples, and structures. However, emotions are a richer psychological phenomenon, the understanding of which helps to treat diseases, control behaviour, and understand the mysteries of the human soul and spirit.

Soul and spirit are the most inaccessible aspects for mastering by digital AI systems. This thesis itself looks like blasphemy against a person since these aspects make a person a person, leaving artificial structures an invariable role of an assistant in achieving their goals. Replacement of a human limb with a prosthesis never makes this prosthesis natural. At the same time, scientists and engineers desire to dive into the depths of the soul and spirit to make AI more useful, perfect, and safe. Of course, various formalisable and non-formalisable interpretations of psychology can help in its implementation.

One can find the concept of "quantum psychology", the content of which is reduced to the mapping of psychological and cognitive phenomena to the processes described in quantum mechanics. Paying tribute to this advanced—quantum—field of knowledge, scientists are trying to build technological analogues or even surpass individual human abilities by using the recommendations of quantum theory. At the same time, no matter what interpretations of quantum physics we take (there are more than ten interpretations), it is worth noting that the essence of quantum computing fundamentally differs from the human cognitive phenomenon. Let's take a few of the aspects.

The quantum superposition. A quantum computer works initially with discrete (binary) data, and the notorious quantum superposition only interpolates data (signal) values at adjacent points (sample values). The accumulation of errors relative to the continuous natural signal accompanies this. The superposition of binary data representation of signals and the collection of errors lead to distortion of the signal spectra that interpret the processes of thinking and mind, obviously detracting from considering such phenomena as those mentioned above continuous human spirit and soul.

The quantum observer principle. It separates the observer from the quantum system. Due to this quantum principle, the system collapse accompanies the moment of observation. In the natural human environment, quite the opposite happens. External observation of a person usually leads to this anxious or excited state, and a persistent observer may even be punished for staring at another person. Upon a doctor's observation, the patient's blood pressure can jump, sometimes even significantly. Therefore, the idea of an observer in quantum and human cases has different manifestations.

The *quantum entanglement* effect is a simultaneous change in the states of two quantum particles that are located from each other over long distances. It's a vanishingly small probability that at least a couple of brain atoms of two different people are in the quantum entanglement state—there are too many atoms in the Universe. Consequently, the non-local aspects of the semantics of AI models must be considered when constructing the semantics of AI models [3].

Psychology has scientific breakthroughs due to applying exact sciences, including physics and mathematics. Developing the tradition of interdisciplinary coupling of various sciences, it is possible to use psychology and AI by introducing the term "*Photonic psychology*". This term has yet to be developed, so let's focus on its potential design features in the context of psychology and AI.

Psychology studies the latent origins and subtle patterns of a person's mental activity and groups of people. This fundamentally defies semantics interpretation in the theory of formal (discrete, digital) systems, which now dominate the AI field. Psychology covers non-formalised aspects of human activity: thoughts, feelings, emotions, sleep, awakening, meditation, etc. The phenomena of psychology cannot be fully represented in discrete, digital, and symbolic formats. Psychology phenomena are continuous—they are more associated with the concept of an electromagnetic field or energy than with a discrete mathematical model, scheme or regulation. Psychological phenomena cannot be accurately represented in metric space.

Psychology operates with concepts that can be described abstractly, such as category theory or topology theory methods. Psychology is a continuous (analogue) phenomenon for mathematicians and physicists. This phenomenon can be well associated with photonic processes, characterised by speed, continuity, and a high rate of signal transformation. Let's consider this continuous psychological phenomenon as an example of a person's manifestation and recognition of emotions using AI systems.

Continuous mathematical and physical features characterise the phenomena of light propagation. Therefore, when building advanced AI systems (general, strong), they may be used to represent human cognitive processes through optical transformation mechanisms. For example, one can imagine a sage's enlightenment process as an instantaneous synergy of an infinite number of coherent rays, solitons or waves in an optical tool—a laser. Such an analogy, by being used in psychological or strategic planning practices together with photonic AI, can make the team's creative activity more successful, help to unravel the competitor's plan, accelerate the achievement of agreement in a team, cure a disease, eradicate fears and stresses, and better understand a person's cognitive abilities.

AI systems can recognise emotions by displaying the result on various, usually discrete, scales with a certain degree of accuracy. Different approaches help to understand discrete emotions such as anger, happiness, anticipation, fear, sadness, and shame. The emotional experience of grief [4] is an example of the close connection between emotion and behaviour. AI systems make the recognition by using classification models, support vector machine algorithms, K-nearest neighbour algorithms, decision tree algorithms, hierarchical type of emotion quantification methods, and deep learning methods.

Nevertheless, the boundaries between the emotional states are fuzzy because the evolutions of emotional states are continuous. A dozen discrete emotional states show only the main aspects of emotion. To cover and demonstrate the whole field of emotions, e.g. the Valence-Arousal Bipolar Coordinate System is proposed [5]. There can be thousands of emotions with an infinite set of shades, which leads to emotional unclearness. In this set, fuzzy boundaries and clear emotional definitions clarify understanding and elaboration on the problems. To diminish the emotional unclearness, the discrete emotions focused on a specific target with relatively intense and short-lived characters are the most studied.

Emotional transparency encourages humans to recognise collective feelings, and dynamic balancing is the human's ability to accelerate agreement in collective decision-making. The collaborative thinking process involves human interactions and deepening inter-understanding during conversations. It can be vividly introduced in the collective strategic planning process [1].

Psychological, behavioural, and social sciences describe the nature of emotion in decision-making processes at the individual and collective levels [6, 7] considering cognitive aspects. The paper [8] suggested three ways of emotional influences in such processes:

1. Non-conscious influence—emotions are a part of the "automatic processes".
2. Emotion regulation—emotion can effectively be managed or adequately displayed.
3. Collective influence—illustrates when a group experiences emotion.

Concerning the first way, a happy leader in urgent situations produces fewer original ideas, but a sad leader makes better decisions [9]; emotion helps to sense threats and opportunities in risky situations; unfair treatment is an antecedent to negative emotions, which leads to a reduction of trust atmosphere [10]. A leader is not fully aware of participants' feelings and cannot quickly recognise the influence of participants' emotions and thoughts. Negative emotions make leaders less prone to take risks and lead to more traditional decisions; on the contrary, positive feelings of leaders encourage them to create new strategies, leading to other positive emotions and high risk-taking [11, 12]. The leader's past emotional experiences may be so intense that team members do not have any possibility to change the leader's stance during a strategic meeting [13].

In the second way, leaders' masking of emotion can stimulate misunderstandings and strategic failure [14]. Leaders' control of emotions eases them to justify a strategic failure [15], and other authority groups' emotional regulation may change their feelings [16]. Representing emotions by a leader shows strategic priority for the team and stimulates a strategic choice [17, 18]; negative leaders' emotions can reduce the team's trust and launch a negative cycle [10]. Negative emotions of an individual have the potential for a long time; in contrast, positive individual emotions can trigger emotional attachment in the direction of identification and integration of the strategic team [19].

Thirdly, strategic activity is a collective process of creating and achieving collective goals, and emotions influence this process by involving the emotional interdependence of participants. Emotions are a collective phenomenon [20]. It may be top-down and bottom-up approaches [21]. The former shows that emotions arise in a whole group, and then, each member joins in the collective emotion; the latter—participants' emotions integrate into collective emotion [22].

The interpersonal transmission of emotions diffuses, mainly via non-conscious and non-verbal ways. Collective emotions may also result from the inclination to share emotions in the strategic group. An example of top-down collective emotions can be threats to identity and trust issues triggered by differences in language, national culture, and professional level of different groups and individuals [23]. Identity threats and lack of trust can lead to disappointment and anxiety and cause collective resistance to strategic changes.

Emotional contagion is a bottom-up process that can occur in informal ways, for example, during small conversations between a meeting's participants and shared with whole groups. Discussions of minor problems can trigger toxic decision-making with negative emotions, which are then distributed among group members [24]. These negative emotions influence strategic group members' understanding of strategic events and behaviour. However, emotional contagion can be a strong driving force for strategic change and improving the situation [25].

Middle managers' perceptions of their leaders' emotions can reduce trust and resistance to strategic change. Fair decisions affect idea- and knowledge-sharing and improve team performance, whereas the opposite leads to anger and obstructive behaviour. Trust can also be achieved as jeopardised through collective emotions depending on whether team members perceive strategic decisions as fair or unfair [26]. A lack of confidence can raise negative emotions in the strategic group, creating fear and strategic myopia.

Emotional control intensifies adverse emotional reactions [7]. It is a way to suppress inadequate emotions in strategic processes to get benefits. During the strategic process, employees may stop displaying negative emotions because such a demonstration could lead to losing decision-making power [27].

The description of emotional factors makes sense only when placed in problematic, spatial, and temporal contexts. Identifying emotions influence a particular collective strategic planning looks like a boundless process in the infinite set of emotions' shades and hundreds of participants' problems. However, the contextualisation of emotion is limited, and understanding the organisational bounds of emotion is in its infancy [8]. Emotions interact with human cognition, but this process cannot be represented in a fully formalised way.

People exchange messages, including voice, text, facial expressions, and gestures. In these processes, different emotional characteristics can be identified. Emotion recognition can be realised based on detecting physical behaviours and physiological activities through voices, neurophysiological activity, videos, texts, gait, etc. Digital statistical and machine learning approaches are usually used to identify emotional states. However, the difficulties of taking into account latent components embedded in the emotion-related data sources, including natural language, facial expressions, speech, body gestures, biosignals, text, and eye gaze, make the accuracy of emotional recognition lower than required. These emotion-related latent components are typically hidden in redundant noises.

Many applications for making emotional recognition, including sentiment analysis tools, exist in different practice fields, such as emotion-aware driver assistance systems for cars, neurology, robot-assisted and music-assisted therapy, emotion-associated information retrieval, enriching user profiles, leisure and entertainment, virtual reality in education, and even emotion recognition in cattle. At the same time, the quality criteria are not satisfied with the restrictions of using the discrete scale of emotions.

Photonic methods of emotion diagnostics and processing make it possible to remove the limitations of digital tools, discrete diagnostics, and pixel recording. For example, they make it possible to replace tens of thousands of binary pixel image

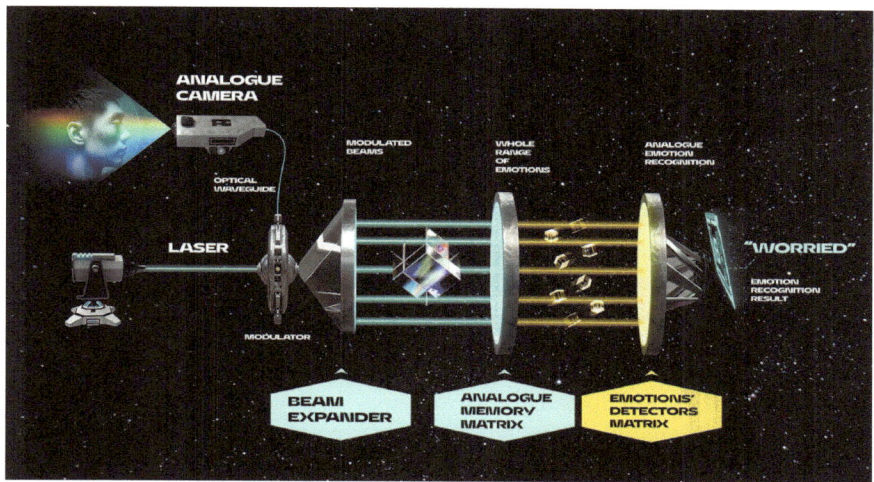

Fig. 3.2 Optical emotion recogniser (published with permission by Openverse Ltd)

comparison operations or millions of Fourier transform operations with a single optical function, representing the path of a light beam through a relatively small optical system (Fig. 3.2, designed using LLM by Openverse Ltd).

In Fig. 3.2, the analogue matrix memory (the emotions' detectors matrix) stores many images, each containing a record of the Fourier convolution of many representations of one of the values of emotions. Such values of emotions can be several thousand. This matrix can be written by training the analogue memory matrix of emotions illustrated in Fig. 10.3 (see Chap. 10).

An analogue (non-digital) camera perceives an external object. It transmits the captured image via a cable by an analogue signal with this image to a modulator, which can be implemented as a transparent film or a solid plate resembling a film photograph. A coherent laser beam passes through the modulator and the beam expander falls on the analogue matrix memory. It is possible to include a preliminary transformation of the modulated beam into a Fourier image. Adequate places of matrix memory resonate with the perceived image, and then information about the result of emotion recognition is taken from the detector matrix.

The concept of analogue recording of an image can be conditioned by obtaining non-digital (continuous) images, for example, on holographic material with its subsequent fixation, multiple readings and storage in memory for a long time without modification (see Chap. 9). The captured image conveys the details of the visible object without its digital transformation.

The advantages of the photonic method of object fixation are the high similarity of the actual object and its continuous image in a wide spectral range of characteristics. Such a photonic approach allows a psychologist (psychoanalyst) to improve the quality of recognition of patients' emotions and the moderator of a strategic meeting to accelerate the achievement of participants' agreement on goals and ways of action.

3.1 Chapter Conclusion

- Psychological practices can be accelerated, and their effectiveness can be increased by AI systems support.
- Digital AI systems can recognise only 7–10 discreet types of emotions; photonic AI can monitor emotions on a continuous scale and cover thousands of shades of emotions.
- Collective strategic meetings can be accelerated by automatically monitoring participants' emotions on a non-discretion scale with photonic AI.
- Methods of planning, mathematics, and physics are used to accelerate psychological procedures; as a result, disciplines such as neuro-linguistic programming, cognitive psychology, and quantum psychology have appeared.
- It is helpful to introduce the term "Photonic psychology", which ensures the synergy of integrating digital and continuous psychological aspects of creating AI systems for accelerating psychological procedures.

References

1. Raikov, A.: Convergent fuzzy cognitive modelling of regional youth policy strategy. In: Yang, X.S., Sherratt, R.S., Dey, N., Joshi, A. (eds) Proceedings of Eighth International Congress on Information and Communication Technology. ICICT 2023. Lecture Notes in Networks and Systems, vol. 694, pp. 911–921. Springer, Singapore (2023). https://doi.org/10.1007/978-981-99-3091-3_74
2. Raikov, A.N.: Accelerating decision-making in transport emergency with artificial intelligence. Adv. Sci. Technol. Eng. Syst. J. 5(6), 520–530 (2020). https://doi.org/10.25046/aj050662
3. Raikov, A.: Cognitive semantics of artificial intelligence: a new perspective. In: Topics: Computational Intelligence XVII. Springer Singapore, Singapore (2021). https://doi.org/10.1007/978-981-33-6750-0
4. Friedrich, E., Wüstenhagen, R.: Leading organizations through the stages of grief: the development of negative emotions over environmental change. Bus. Soc. 56, 186–213 (2017)
5. Langeslag, S.J.E.: Effects of organization and disorganization on pleasantness, calmness, and the frontal negativity in the event-related potential. PLoS ONE 13(8), e0202726 (2018). https://doi.org/10.1371/journal.pone.0202726
6. Hodgkinson, G.P., Healey, M.P.: Psychological foundations of dynamic capabilities: reflexion and reflection in strategic management. Strat. Manag. J. 32, 1500–1516 (2011)
7. Huy, Q.N., Guo, Y.: Middle managers' emotion management in the strategy process. In: Floyd, S.W., Wooldridge, B. (eds.) Handbook of Middle Management Strategy Process Research, pp. 133–153. Edward Elgar Publishing, Cheltenham (2017)
8. Brundin, E., Liu, F., Cyron, T.: Emotion in strategic management: a review and future research agenda. Long Range Planning 55(4), 102144 (2022). https://doi.org/10.1016/j.lrp.2021.102144
9. Treffers, T., Klarner, P., Huy, Q.N.: Emotions, time, and strategy: the effects of happiness and sadness on strategic decision-making under time constraints. Long Range Plan. 53, 101954 (2020)
10. Smollan, R.K.: Trust in change managers: the role of affect. J. Organ. Change Manag. 26, 725–747 (2012)
11. Delgado-García, J.B., De La Fuente-Sabaté, J.M., De Quevedo-Puente, E.: Too negative to take risks? The effect of the CEO's emotional traits on firm risk. Br. J. Manag. 21, 313–326 (2010)

12. Håkonsson, D.D., Eskildsen, J.K., Argote, L., Monster, D., Burton, R.M., Obel, B.: Exploration versus exploitation: emotions and performance as antecedents and consequences of team decisions. Strat. Manag. J. **37**, 985–1001 (2016)
13. Kisfalvi, V., Pitcher, P.: Doing what feels right—the influence of CEO character and emotions on top management team dynamics. J. Manage. Inquir. **12**, 42–66 (2003)
14. Vuori, N., Vuori, T.O., Huy, Q.N.: Emotional practices: how masking negative emotions impacts the post-acquisition integration process. Strat. Manag. J. **39**, 859–893 (2018)
15. Vince, R.: Being taken over: managers' emotions and rationalizations during a company takeover. J. Manag. Stud. **43**, 343–365 (2006)
16. Vuori, T.O., Huy, Q.N.: Regulating top managers' emotions during strategy making: Nokia's socially distributed approach enabling radical change from mobile phones to networks in 2007–2013. Acad. Manag. J. (2021). https://doi.org/10.5465/amj.2019.0865
17. Liu, F., Maitlis, S.: Emotional dynamics and strategizing processes: a study of strategic conversations in top team meetings. J. Manag. Stud. **51**, 202–234 (2014)
18. Loewenstein, G.: Emotions in economic theory and economic behavior. Am. Econ. Rev. **90**, 426–432 (2000)
19. Raitis, J., Harikkala-Laihinen, R., Hassett, M., Nummela, N.: Finding positivity during a major organizational change: in search of triggers of employees' positive perceptions and feelings. Emotions and Identity **13**, 3–16 (2017)
20. Barsade, S.G., Gibson, D.E.: Why does affect matter in organizations? Acad. Manag. Perspect. **21**, 36–59 (2007)
21. Raikov, A., Pirani, M.: Human-machine duality: what's next in cognitive aspects of artificial intelligence? IEEE Access. **10**, 56296–56315 (2022). https://doi.org/10.1109/access.2022.3177657
22. Barsade, S.G., Gibson, D.E.: Group emotion: a view from top and bottom. Res. Manag. Groups Teams **1**, 81–102 (1998)
23. Huy, Q.N.: How middle managers' group-focus emotions and social identities influence strategy implementation. Strat. Manag. J. **32**, 1387–1410 (2011)
24. Maitlis, S., Ozcelik, H.: Toxic decision processes: a study of emotion and organizational decision making. Organ. Sci. **15**, 375–393 (2004)
25. Erkama, N., Vaara, E.: Struggles over legitimacy in global organizational restructuring: a rhetorical perspective on legitimation strategies and dynamics in a shutdown case. Organ. Stud. **31**, 813–839 (2010)
26. Kim, W.C., Mauborgne, R.: Procedural justice, strategic decision making, and the knowledge economy. Strat. Manag. J. **19**, 323–338 (1998)
27. Bryant, M., Wolfram Cox, J.R.: The expression of suppression: loss and emotional labour in narratives of organisational change. J. Manag. Organ. **12**, 116–130 (2006)

Chapter 4
Situational Emotions

The collective decision-making process can be accelerated by considering participants' emotional dynamics. This chapter addresses digital emotion recognition methods, which can identify only some discrete emotions. Photon AI diagnostics can remove this limitation and make emotional recognition on a continuous scale.

4.1 Strategic Emotions

The collective decision-making process can be accelerated crucially by using situational centres (SC), which help a team analyse complex problems comprehensively, purposefully, and in a stable way [1]. The SC's process is immersed in a hybrid (human–machine) environment. The existing limitations of SC are as follows: there is a skew in the direction of solving analytical tasks to the detriment of synthesis tasks, remote experts cannot understand each other quickly, and the discussion process may be too divergent.

Interactive questionnaires, the convergent meeting procedure, the hierarchical order of goal setting, the situational awareness framework, etc., are used [2] to speed up the collective decision-making process. Convergent technology is associated with the inverse problem-solving approach, which prescribes the order of organising decision-making processes and uniquely structuring information.

Multidimensional visualisation, virtual reality, graphics, semantic maps, etc., are also used. The following aspects are considered: self-organising interactions, assembling the strategic actors, network expertise, collective intelligence methods, evolutionary decision-making methods, rationalising features of human thinking, augmented reality, etc. The decision-making meeting for creating a strategy draft for a branch of the country's region economy may take 2–3 days [3, 4].

© The Author(s), under exclusive license to Springer Nature Singapore Pte Ltd. 2024   33
A. Raikov, *Photonic Artificial Intelligence*,
SpringerBriefs in Computational Intelligence,
https://doi.org/10.1007/978-981-97-1291-5_4

The decision-making meeting may be accelerated by monitoring participants' emotions, including contagion of the mood of pessimism, negativity, and confrontation between the decision-makers when facing challenging, complex, and empathy-arousing events. Emotions can be supportive or obstructive in collective decision-making. Emotion recognition methods using digital computers and cameras help identify some discrete emotions. Unfortunately, emotions are not easy to foresee or identify; they can change quickly, and subtle clues should be sensitively captured and utilised to infer emotions. There can be thousands of emotions, and the boundaries between different emotions are unclear.

However, in digital computers, to diminish unclearness, the discrete emotions focused on a specific target or cause with relatively intense and short-lived character are the most studied. Strategy processes may discretely include anger, happiness, sadness, fear, shame, guilt, commitment, etc. The emotional experience of grief [5] is an example of the close connection between emotion and strategic management. Emotional transparency encourages a leader to recognise collective feelings, and dynamic balancing is the leader's ability to accelerate getting an agreement in collective decision-making.

There are significant restrictions in finding the connection between emotions and analytical thinking. Emotions are seen as umbrella constructs only for describing affective phenomena without detailed defining the differences between discrete emotions, affect, or moods. The restrictions relate to the technological issue—the same emotion constructs and digital machine learning (ML) are used to recognise different emotional phenomena. A significant problem is emotional messiness, which is becoming a cause of confusion and confounding effects in decision-making. Behavioural and social sciences describe the nature of emotion in decision-making processes at the individual and collective levels [6].

The description of emotional factors makes sense only when placed in problematic, spatial, and temporal contexts. However, the digital contextualisation of emotion is limited. Emotions interact with human cognition, but this process cannot be fully represented in a formalised (digital) way. The representation of emotions has a continuous, non-digital character.

4.2 Emotions Recognition by Brain Signal or Face

Participants of networked meetings exchange messages, including represented by voice, text, facial expressions, body gestures, eye gaze, etc., which can be used to identify different emotional characteristics. Emotions, in turn, are a psychological phenomenon reflected in communicated acts, physical behaviours, and physiological activities. Emotion recognition can be realised by automatically detecting physical behaviours and physiological activities.

Statistical and ML approaches of AI systems can identify emotional states in real time. However, the difficulties of considering latent components embedded in the emotion-related digital data sources make the accuracy of emotional recognition by digital AI systems lower than required. These emotion-related latent components are typically hidden in various redundant noises.

There are many AI applications and tools for emotional recognition, including sentiment analysis, in different practice fields: emotion-aware driver assistance systems for cars, neurology, robot-assisted and music-assisted therapy, emotion-associated information retrieval, enriching user profiles, leisure and entertainment, virtual reality (VR) in education, and even emotion recognition in cattle, etc.

One of the main approaches in studying emotion recognition processes is utilising brain activity information by detecting brain signals from electroencephalogram (EEG) emotion recognition tools [7]. Psychophysiology-based approaches complement facial or speech information recognition methods in studying the emotional cognition mechanism, mental workload, fatigue, falling asleep, motor imagery, attention, sincerity of statements, etc.

It is not a comforting way to attach the EEG system's electrodes to participants of the strategic meeting. When VR is widely used, the participants can attend the meeting online, and their physiological metrics can be monitored if the VR equipment is furnished with various sensors. The EEG approach helps to understand discrete emotions such as anger, anticipation, fear, and sadness. A dozen discrete emotional states show only the main aspects of emotion. To cover the entire spectrum of emotions, e.g. the Valence-Arousal Bipolar Coordinate System is used [8].

There are two different views on the connection of neural sources with emotions. The first one, local, insists that some discrete emotions reflect the brain's unique structure. The second one, distributed—is that not only a single anatomical structure uniquely specialises in emotions, but that the human emotion is a product of the cooperation of multiple cortex regions [9, 10].

Neural responses to facial expressions can occur with latencies of 100 ms [11], and only two eye fixations are sufficient to recognise facial emotions [12]. Emotion recognition can rely on individuals, on multiple facial features, or be more holistic, which implies perceptual integration across the whole face. Specific face regions are relevant for decoding emotions from facial expressions.

The mouth is the most important for discriminating basic facial expressions; it helps recognise happy expressions from the bottom rather than the top of the face. Faces may be processed holistically [13].

4.3 Emotions Recognition by Text

Several ML approaches help to recognise emotions from the text created during conversations [14, 15]. It may be Naive Bayes, K-nearest neighbours algorithms, generalised linear model, fast-large margin, bidirectional long short-term memory, convolutional neural network (CNN), decision tree, recurrent neural network (RNN), random forest, support vector machine, term frequency, word2vec which can be used to convert the set of words into a vector space. ML models can provide an accuracy score—of 0.92, recall—of 0.902, and precision—of 0.902 [14]. One of the essential reasons for losing the quality is that words have different meanings in different contexts. The decision tree and random forest are unsuitable for getting such results with the dataset based on the effective tweets [16]—about 7000 emotions labelled utterances [14].

All digital emotional recognition methods can be divided into three classes: pattern-matching, ML, and deep learning. These methods can be used to recognise participants' intentions using the text query format. Nevertheless, the meanings of words are polysemous, which can cause inaccurate emotion recognition. The accuracy of emotion classification can be up to 95%. The general emotion extraction algorithm includes data collection, pre-processing and emotion annotation, feature extraction, classification, and actionable pattern discovery. It is essential to ensure that the balance between extracted continuous contextual and fine-grained (discrete) information is generated during a strategic dialogue. Contextual information can be extracted from the relevant images and texts, including historical ones and digital data of the messages' texts. It may conflict between features extracted at the context level and the digital one.

The paper [17] suggests combining the interaction of hierarchical feature-related information, which supports recognising dialogue emotion and dialogue act/intent separately and then analysing the cross-impact of these two components. It helps to smooth out the conflict between fine-grained hierarchical functions and functions at the context level by obtaining deep semantic information through interactively extracting emotional and behavioural characteristics with ML. The paper proposes a hierarchical feature interactive fusion network method that enriches semantic information using emotion and act/intent features in dialogue. The accuracy of emotion recognition was 49%, and the accuracy of act/intent recognition was 82%. It was shown that among seven known emotions, neutral is the best for recognition, followed by anger, disgust, and happiness. However, fear, sadness, and surprise have low accuracy in recognition.

The dialogue may be as follows: chat with emotional perception [18], persuasive discussion [19], and visual question and answer [20]. The technologies need the help of synonym sets and emotion dictionaries. The paper [21] proposed the system DialogueRNN, which tracks participants' state and emotional changes by considering contextual information. DialogueGCN system [22] uses a graph convolution neural network to capture emotions in short conversations. This and other neural network methods are sensitive to emotional changes in continuous dialogue but are restricted by non-high-accurate discrete scales of emotions.

4.4 Compressing Meeting Time by Considering Emotions

In general, collective strategic meeting technologies are based on the methods of strategic analysis and networked expertise [3]. For example, these technologies can include such procedures as gathering experts' comments, estimating events by differential scales, monitoring problem situations, electronic brainstorming, planned networked problem-solving (up to 25 participants), networked strategic congress (up to 250 participants), etc.

Electronic brainstorming is the most challenging group decision-making process, which may be divergent or convergent. This process must be convergent to get an agreement between participants in a limited period. The author's convergent approach can be used for it [2]. When the problem is incorrectly framed, the author's convergent approach guarantees the necessary structural conditions for stable and purposeful decision-making.

Semantic interpretations of text messages generated during brainstorming and virtual collaborating tools help to accelerate the process. These tools support electronic brainstorming by dynamically demonstrating on the screen the characteristics of participants' speeches, such as talking time, loudness, and voice frequencies, in real time. However, the strategic meetings are sometimes too long, which is unacceptable in an emergency [2]. The idea of further speeding up meetings is to consider the possibility of emotional recognition. For this, the architecture of SC has to be added by the emotional recognition subsystem, as illustrated in Fig. 4.1.

The emotional recognition subsystem of SC consists of two main parts: training and decision. The first part—the upper level of Fig. 4.1—addresses preparing the keywords (factors, concepts) and fragments of cognitive models equipped with trained neural networks for emotion recognition. It is better to consider the different individual features of the strategic team's participants: skills, age, place of birth, cultural traditions, publications, etc. Experts may create the origin keywords and model fragments. The result of cognitive modelling can be verified or added by mapping ones on the relevant big data [23]. An emotional pattern by ML supplies

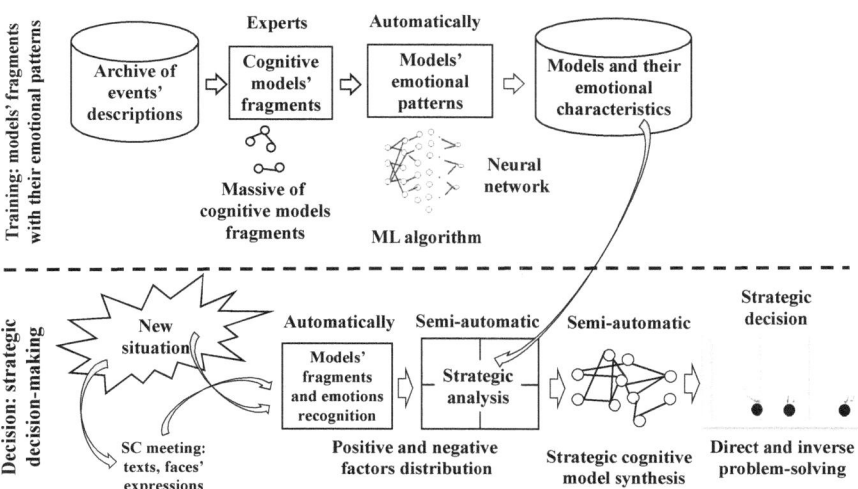

Fig. 4.1 Emotional recognition subsystem of SC

every model's fragment. One of the mentioned emotion recognition technologies can be separately supplied in this part.

The second part of the subsystem—the lower level of Fig. 4.1—supplies team strategic meetings by automatically creating suggestions to include some essential candidates of keywords or models' fragments in the strategic analysis process. This part of the subsystem analyses participants' messages, statements, and facial expressions to ensure the automatic recognition of their emotions. These emotions, along with the connections to participants, are attributed to the words expressed in such a dynamic and synchronised way that meeting participants can watch them on the collective screen of SC or remote participants's computer screens.

4.5 Implementations

The authors' experience of moderating strategic meetings in SC with the participation of 7 to 35 participants shows that they often hide their emotions. Currently, only the professionalism of the meeting moderator can help identify the participants' emotional state. Thus, in particular, the moderator can understand how sincere the participants' statements are. Facial expressions, gaze, and movement of hands help to identify the emotions.

Let us consider the process of determining and demonstrating in real time on the collective screen of SC the characteristics of messages generated by the meeting participants, which supports self-regulation of the participants' activity. There can be two extremes: one participant can talk for a long time and write long messages, and the other can be silent. Observing this demonstration makes the first one less active and the second more energetic. A separate task was the automatic assessment of the proximity of the participants' positions during their statements on the meeting topic. To do this, the index of the polarisation of opinions ρ_{PO} and the parameter of the deviation of the semantics of the utterance from the meeting topic d_{SD} are introduced and calculated. This made it possible to display the discussion process in dynamics on the SC's collective screen using a two-dimensional coordinate system as a "Hodograph of opinions" (Fig. 4.2).

The working area of the phase plane (Fig. 4.2, the top left) is divided into four non-intersecting rectangular zones: "nominal", "operating deviations", "marginal deviations", and "emergency". The moment at the end of a message is marked with a white disc; when somebody hovers the cursor over the disc, a hint indicates the message. Three parameters are entered for each message: τ_M—the duration of writing a message, measured in seconds; v_M—the volume of the message, which is a convolution of two quantities: the length of the message, expressed in the number of words and the size of the message, described in the number of characters; v_{MT}—the volume of the message, normalised by the ratio of the number of proposals in the message on the subject of the meeting to the total number of recommendations in the message.

The diagram shown in Fig. 4.2 below is based on the three entered parameters. Each block's colour and dimensions make sense of the three parameters listed. Thus,

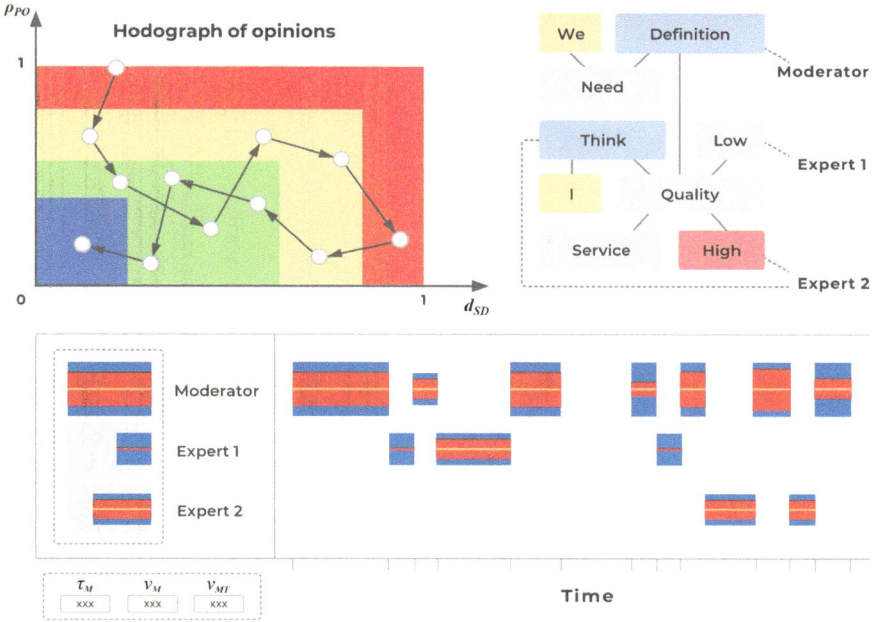

Fig. 4.2 Hodograph of the opinions dynamic of the meeting participants

such a presentation is cognitive and allows participants to visualise each participant' contribution in the meeting to the discussion process and show its effectiveness concerning the topic of the meeting.

Considering the subject of discussion, a meaningful analysis of messages allows building a graph of the core semantic relations between critical concepts (Fig. 4.2, at the top right). It is also posted, for example, by assessments of participants' opinions on the subject of discussion.

4.6   Chapter Conclusion

- Digital emotion recognition approaches can only give discrete results which creates the requirements for emotion diagnostics on a continuous scale that can be done with a photonic approach.
- There are dozens of approaches to understanding strategic thinking that help speed up collective strategic meetings; however, monitoring participants' emotions using AI tools is comparatively new.
- A unique SC subsystem must be implemented to cover the room in emotion recognition, which can estimate the emotional meanings of the strategic conversation discussion entities and map them to the screens.

- The potential of emotion recognition by computer vision, ML, and EEG tools in their analogue and digital versions must be integrated to accelerate strategic meetings.

References

1. Ilyin, N., Malinetsky, G., Kolin, K., Zatsarinny, A., Raikov, A., Lepskiy, V., Slavin, B.: Distributed situational centres system of cutting-edge development. In: Proceedings of 10th International Conference on Management of Large-Scale System Development (2017). https://doi.org/10.1109/mlsd.2017.8109638
2. Raikov, A.N.: Accelerating decision-making in transport emergency with artificial intelligence. Adv. Sci. Technol. Eng. Syst. J. **5**(6), 520–530 (2020). https://doi.org/10.25046/aj050662
3. Raikov, A.: Convergent fuzzy cognitive modelling of regional youth policy strategy. In: Yang, X.S., Sherratt, R.S., Dey, N., Joshi, A. (eds.) Proceedings of Eighth International Congress on Information and Communication Technology. ICICT 2023. Lecture Notes in Networks and Systems, vol. 694, pp. 911–921. Springer, Singapore (2023). https://doi.org/10.1007/978-981-99-3091-3_74
4. Gubanov, D., Korgin, N., Novikov, D., Raikov, A.: E-expertise: modern collective intelligence. In: Series: Studies in Computational Intelligence, vol. 558, p. XVIII. Springer, Cham (2014). https://doi.org/10.1007/978-3-319-06770-4
5. Friedrich, E., Wüstenhagen, R.: Leading organizations through the stages of grief: the development of negative emotions over environmental change. Bus. Soc. **56**, 186–213 (2017)
6. Huy, Q.N., Guo, Y.: Middle managers' emotion management in the strategy process. In: Floyd, S.W., Wooldridge, B. (eds.) Handbook of Middle Management Strategy Process Research, pp. 133–153. Edward Elgar Publishing, Cheltenham (2017)
7. Li, X., Zhang, Y., et al.: EEG based emotion recognition: a tutorial and review. ACM Comput. Surv. **55**, 1–57 (2022). https://doi.org/10.1145/3524499
8. Langeslag, S.J.E.: Effects of organization and disorganization on pleasantness, calmness, and the frontal negativity in the event-related potential. PLoS ONE **13**(8), e0202726 (2018). https://doi.org/10.1371/journal.pone.0202726
9. Britton, J.C., Phan, K.L., Taylor, S.F., Welsh, R.C., Berridge, K.C., Liberzon, I.: Neural correlates of social and nonsocial emotions: an fMRI study. Neuroimage **31**(1), 397–409 (2006)
10. Tang, J., LeBel, A., Jain, S., et al.: Semantic reconstruction of continuous language from non-invasive brain recordings. Nat. Neurosci. **26**, 858–866 (2023). https://doi.org/10.1038/s41593-023-01304-9
11. Ramdani, C., Ogier, M., Coutrot, A.: Communicating and reading emotion with masked faces in the Covid era: a short review of the literature. Psychiatry Res. **316**, 114755 (2022). https://doi.org/10.1016/j.psychres.2022.114755
12. Schurgin, M.W., Nelson, J., Iida, S., Ohira, H., Chiao, J.Y., Franconeri, S.L.: Eye movements during emotion recognition in faces. J. Vis. **14**(13), 14 (2014). https://doi.org/10.1167/14.13.14
13. Kilpeläinen, M., Salmela, V.: Perceived emotional expressions of composite faces. PLoS ONE **15**(3), e0230039 (2020). https://doi.org/10.1371/journal.pone.0230039
14. Chowanda, A., Sutoyo, R., Meiliana, Tanachutiwat, S.: Exploring text-based emotions recognition machine learning techniques on social media conversation. Proc. Comput. Sci. **179**, 821–828 (2021). https://doi.org/10.1016/j.procs.2021.01.099
15. Malova, I.S., Tikhomirova, D.V.: Recognition of emotions in verbal messages based on neural networks. Proc. Comput. Sci. **190**, 560–563 (2021). https://doi.org/10.1016/j.procs.2021.06.065
16. Bravo-Marquez, F., Frank, E., Pfahringer, B., Mohammad, S.M.: Affectivetweets: a weka package for analyzing affect in tweets. J. Mach. Learn. Res. **20**, 1–6 (2019)

17. Gan, C., Yang, Y., Zhu, Q., Jain, D.K., Struc, V.: DHF-Net: a hierarchical feature interactive fusion network for dialogue emotion recognition. Expert Syst. Appl. **210**, 118525 (2022). https://doi.org/10.1016/j.eswa.2022.118525
18. Zhou, L., Gao, J., Li, D., Shum, H.Y.: The design and implementation of XiaoIce, an empathetic social chatbot. Comput. Linguist. **46**, 53–93 (2020)
19. Chen, H., Ghosal, D., Majumder, N., Hussain, A., Poria, S.: Persuasive dialogue understanding: The baselines and negative results. Neurocomputing **431**, 47–56 (2021)
20. Ye, Y., Zhang, S., Li, Y., Qian, X., Tang, S., Pu, S., Xiao, J.: Video question answering via grounded cross-attention network learning. Inf. Process. Manage. **57**, 102265 (2021)
21. Majumder, N., Poria, S., Hazarika, D., Mihalcea, R., Gelbukh, A., Cambria, E.: DialogueRNN: An attentive RNN for emotion detection in conversations, pp. 6818–6825. Proceedings of the AAAI conference on artificial intelligence, AAAI (2019)
22. Ghosal, D., Majumder, N., Poria, S., Chhaya, N., Gelbukh, A.: DialogueGCN: a graph convolutional neural network for emotion recognition in conversation. In: Proceedings of the 2019 conference on empirical methods in natural language processing and the 9th international joint conference on natural language processing, EMNLP-IJCNLP, pp. 154–164 (2019).
23. Raikov, A., Ermakov, A., Merkulov, A., Panfilov, S.: Automatic synthesis of cognitive model for revealing economic sectors' needs in digital technologies. In: Yang, X.S., Sherratt, S., Dey, N., Joshi, A. (eds) Proceedings of Seventh International Congress on Information and Communication Technology. Lecture Notes in Networks and Systems, vol. 448. Springer, Singapore (2023). https://doi.org/10.1007/978-981-19-1610-6_20

Chapter 5
Photonic Thought

In Artificial Intelligence (AI), "Photonic thought" is better than a digital one representing natural human thought because the former considers the analogue signals and behaviour of the atomic level of the human neural network. Digital neural networks are trained for a long time and require high energy costs. Human thought can solve a complex problem with much less energy and literally with the speed of light.

Since ancient times, it has been believed that if thoughts are deep and meaningful, they cannot be fully expressed in words. Words expressing feelings and thoughts make up a small part of the solution. The true meaning and purpose of the solution are in infinite depth. Words seem to protect the meaning of the decision—the true meaning and purpose lie in the zone of silence.

The solution is invisible; its information shell is visible—words, drawings, and diagrams. But all this is only data that carries information, partially revealing its true meaning. The real meaning of information is behind the veil of words.

It is only possible to understand the meaning of the decision brought by words if one feels the whole situation. Words swaddle the solution, hiding its true beauty or ugliness. Words carry information, the meaning of which is still worth unravelling. Actions which words initiate can reveal this meaning.

Words provide an information service, which partially stimulates thoughts and causes feelings. An inexpensive information service can excite a beneficial thought, and an expensive one can mislead a person. At the same time, if the words are misunderstood, they can cause inadequate feelings.

The words in the decision may be like dust in the eyes: there are eyes, but they see nothing. The meaning is soft, the word is solid, and the word can destroy the meaning of human intention and decision. Words can lie; facial expressions and gestures—never. A shot of a glance is more vital than a word. It is only necessary to recognise these manifestations of the organism, which have infinite depth and wave spectrum.

Are AI systems capable of recognising the meaning of words, having a discrete form of information representation brought by a digital signal with a limited spectrum?

Currently, digital machine learning tools require significant memory, which exceeds the capacity of a single or multiple computing unit (CU)—deep learning accelerators. Distributed deep learning systems are being created for this, which relies on parallelisation to place a single training task on multiple CUs. Thousands of CU can be connected in a single group or by inter-group communication. However, current hardware can provide only high-bandwidth connections for a limited group of CUs because increasing communication speed in the single group has not been accompanied by the same increase in inter-group bandwidth. This discrepancy limits the communication efficiency during the training of neural networks [1].

This communication limitation can be overcome by applying optical tools [2]. Light used to connect computing nodes can enormously decrease computer energy consumption while increasing bandwidth. For example, in microresonator Kerr frequency combs, independent information channels can be encoded onto many light colours for massively parallel data transmission with low energy consumption [3].

Data and model parallelisms are significant for digital neural network systems. In the first case, each CU keeps a copy of the entire training model; in the second case, each CU keeps a full copy of the dataset and receives a training model of every section. It can be horizontally or vertically parallelisms (pipeline and tensor). Collective operations tools synchronise the stages of each of these parallelism strategies.

For the Artificial Mind (see Chap. 2) hardware representation, the digital industry's discrete semiconductor basis can be supplemented with optics (photonic) tools that process continuous signals without cutting up their natural infinite-dimensional spectrum [4]. Optics already allows mathematical operations, such as convolution, differentiation and integration of functions, Fresnel transformation, scaling of function arguments, and restoration of functions from the spectral density of the sum of this function with the δ-function. Holographic principles are used in these operations.

Many of these operations, including operations of complex multiplication or integral transformations implemented in optical systems, are based on the modulation of light passing through the transparent screen. The effect of diffraction of light waves is used for modulation. Photonic technologies can also extract or insert specific wavelengths from or into some set using a resonator, resulting in a compact wavelength-selective routing operation. It results in an all-to-all topology with reduced component count. For this implementation, paper [2] suggests the silicon photonic accelerated compute cluster architecture, which embeds photonic transceivers, high-bandwidth links, and a multi-wavelength selective switch that maps data flow to wavelengths. The testbed experiment used a frequency comb source where a cascaded ring switch shuffles an array of wavelengths. Paper [5] suggested splitting and combining wavelength channels to scale the data transmission bandwidth. Papers [6, 7] suggested the optical circuit switches to reconfigure the network topology dynamically for different traffic demands. The experiments demonstrated the feasibility of the proposed architecture. For example, it was shown that due to the suggested optical architecture on four-GPU, a 22% performance improvement relative to a similarly sized traditional topology can be achieved [2].

There have been many attempts to create optical neural networks. The paper [8] demonstrates the all-analogue photoelectronic chip for high-speed vision tasks

(ACCEL)—optics are implemented in more than 99% of them. It integrates diffractive optical analogue computing and electronic analogue computing with a high level of scalability, nonlinearity, and flexibility. Experimental tests conducted on ACCEL outperform leading digital GPUs by processing high-resolution images of scenes more than 3,000 times faster while consuming 4,000,000 times less energy. It achieves classification accuracies of more than 85% over 100 testing samples.

However, according to ACCEL's authors, this system needs to add more layers of optical parts and re-design electronic components for parallel outputs with more sensitive photodiode arrays. It also needs more complicated network structures with a larger size or several ACCELs can be cascaded. As the ACCEL require coherent light sources, an active detection has to be created. This system uses bit-semiconductor electronic components with a chip area of about 2×2 mm^2. This means its work is based on analogy with a digital information processing algorithm and, as a result, will be accompanied by discrete distortion of the signal spectrum. The system does not use the Fourier transform of images, which does not allow for a reduction of the system's physical size. The authors do not make the statement about rewritable memory or consider the non-local quantum effects, which can play an essential role in creating and processing the cognitive semantics of AI models (see [4]).

The paper [9] suggests a fully integrated coherent optical neural network with weight-stationary architecture. This architecture ensures a silicon photonic foundry process incorporating edge couplers, phase shifters, waveguide-integrated germanium photodiodes, and carrier-based microring modulators. The weight-stationary architectures compute a matrix–vector product per cycle [10, 11], which are well-suited for applications that require processing data with ultra-low latency. A bottleneck of this neural network is training on a large training set to optimise weight parameters. Such applications allow real-time inference, training directly on optical signals, and avoiding electrical-to-optical conversions.

For further accelerating computations, one must reject the data's binary (digital) representation and operate continuous data by exploiting optical tools. Thus, the digital representation of data can be replaced by a continuous (analogue) one. However, such a substitution of digital representation data cannot be absolute since optical transformations have a corpuscular-wave nature. A photon with zero rest mass and spin 1 carries no electric charge. The quantum mechanical aspect of light is characterised by the wave function, which remains the same under an interchange of two photons: they are bosons and obey *the Bose–Einstein statistics*.

The theory of autonomous elementary computational operations on the optic-holographic basis was developed long ago. The possible architecture of advanced optical technological decisions for these conditions can be a holographic memory (HM). For example, paper [12] suggested the hybrid holographic correlator architecture for detecting image matches in a shift, scale, and rotation-invariant manner. The system in [4] (see also Chap. 10) consists of a laser, deflectors, coherent signal converters, and quickly rewritable HM. The Fourier transforms also can be adopted as a methodological basis for optical transformations.

Creating such HM requires considering the phenomenon partially resembles "quantum gravitation" effects, which may play a role in the nanoscale. A fundamental theory of quantum gravity might describe the Universe and the matter's particle behaviour; however, there is no such theory yet. The computational properties of holographic materials in the context of quantum field theory and the idea of quantum gravity might be found [13]. The gravitational anti-de-Sitter space ("the bulk") and non-gravitational description as the conformal field theory ("the boundary") somebodies are applied for these aims.

The information processing of the bulk-boundary space–time of the materials may have a causal structure. Bulk and boundary causal mechanisms can take different forms, constraining computational processing. For example, the quantum non-local effect (the entanglement) can influence computations in a finite bulk region. For special bulk regions called scattering regions, computations are reproduced on the boundary by the entangled system.

Creating holographic materials involves finding an approach to ensure many stable states of bulk's quantum particles in the context of writing and reading information. For this aim, the bulk quantity can be assessed using boundary degrees of freedom of the quantum mechanical system (described by a density matrix ρ) and the von Neumann entropy method that is:

$$S(\rho) = -\mathrm{tr}(\rho \log_2 \rho), \tag{5.1}$$

where tr is the trace and $\log_2$ is the matrix logarithm.

In conformal field theory, different states of matter can be calculated using the Hubney-Ryu-Takayanagi proposal for entanglement entropy to measure the correlation of quantum states [14]. This proposal is a hypothesis in holography theory that establishes a quantitative relationship between the entropy of entanglement of bulk elements and the geometry of the associated anti-de Sitter boundary space–time.

The scattering region into the bulk can take the shape of a tetrahedron with curved sides. The inputs and computational processes may change the bulk geometry and stability. The boundary description is the unique means for understanding processes that can happen in bulk. However, boundary description cannot precisely capture the quantum processes inside the scattering region of bulk. Besides knowing direct inputs and outputs that appear near the boundary, the bulk process can involve messages sent from different internal parts of the bulk and outside the bulk by entanglement effect.

Quantum complexity is the time and space resources needed to efficiently write, read, store, and compute processes in HM. It is the minimum number of elementary operations that the holographic material must ensure. The representation of complexity as the entropy of a system with unitary operators is known as the thermodynamics of complexity [15]. The behaviour of holographic subregion complexity near the critical point may be elaborated by quantum field theory with a crucial point [16].

The site [17] advertises a fully optical neural network architecture to enhance computational speed and power efficiency over traditional digital computers. They suggested a model of recurrent neural networks (RNN) that combines the remembering ability of RNNs with the ability to forget irrelevant information. This artificial neural network can help to solve the gradient explosion/vanishing problem and enable neural networks to learn long-term correlations in the data.

Thus, previous optical neuron networks based on matrix multiplication have manufacturing limitations, resulting in limited scalability. As shown, an optical signal has a great potential for creating neuron networks, which are characterised by low power consumption, low latency, wide bandwidth, and high parallelism. It is critical to scale up the dimensions of matrix size and neuron networks (depth, width, number of layers, and neurons in a layer) [18]. However, current results show that the compute rate and matrix size are beyond the promised aims [19, 20]. Modern optical hardware supports only small matrices and few neurons. To divert these limitations, the sampling operations are used before photonic computing [21], accompanied by the limitations inherent in those mentioned above for processing digital information. The temporal, wavelength, and spatial dimensions with Mach–Zehnder modulators can help to perform large-scale matrix computation [22].

The significance issue arises: *is there material with multiple stable quantum states that a modulated laser beam can quickly change?*

Such photonic material must be created for *optical deflection* and *quickly rewritable 3D holographic memory*. A finite number of images must be rapidly written and re-written by laser beams in one location ("point", "spot", or "dot") of the holographic memory. Some materials, such as lithium niobate ($LiNbO_3$) and phenanthrenequinone-doped poly (methyl methacrylate), can ensure such things. Still, they can provide writing only one image in a single point, which changes when a laser beam reads it. The photon is stable, but the material for recording is unstable relative to the photon. However, it must be multi-stable on the group of atoms and quantum structure level. The photonic material must also have transparency properties to serve as a signal modulator.

Materials specialists use databases with structures of materials and their characteristics and evolutionary computing programs for synthesising materials. Without adequate computing power, algorithms are reduced. For example, the Schrodinger equation is replaced by generalised calculations based on energy parameters using the density functional theory (DFT) methods.

Currently, the central part of the scientific work in creating new materials is devoted to synthesising harder and more temperature-resistant materials and new drugs, assessing the activity of organic compounds, etc. For this, databases of material structures are used; for example, there are well-known organic crystal structure databases where several hundred thousand structures are stored. Laws of quantum physics, equations of molecular dynamics, the Periodic Table, etc., are also used.

The material synthesising processes are usually too complex for supercomputer calculations. It may require more than 50^{100} operations to calculate diffraction to find atomic positions in a crystal. The accuracy of the parametrisation limits such approaches as the Lennard–Jones potentials and van der Waals interactions [23] because the structure is complicated: surface atoms experience a different electrostatic field, bulk-derived potentials may fail in describing molecules, and the atomic environment strongly differs in a natural and perfect crystal.

Even for non-complex systems, interatomic potentials may be inappropriate for finding out classically by first-principles methods—all the ways that are based upon the determination from the fundamental theorems of quantum mechanics. Numerical simulations can help to explain the experimental data. Hartree and Fock's program [24] proposed a solution to such a problem by simplifying assumptions about the wave function.

The HM system can be designed as a neural processor. The core of this device is an optical neural network constructed as a sequence of matrix blocks of 3D holographic rewritable memory containing light-transparent dots with recorded data, which are used for the modulation of beams. For such dots, the output brightness is proportional to the brightness of the incoming signals and depends on the material's transparency. The output beam from every dot is split by the matrix of deflectors and directed to the many (maybe all) dots of the following matrix block (see Chap. 10).

Written images in every dot of the holographic image have to be changeable, which must be ensured by optical methods and tools. It is necessary to develop optical fibres, deflectors, and rewritable holographic lenses whose light characteristics (conductivity, directions of light) can be changed by external (e.g. laser) influence. Therefore, a neural network is trained by changing these dots' transparent (modulated by images) characteristics. The device input can be a "laser array" connected to holographic transducers of beams. This can be a coherent radiation source, a beam expander, polarisers, deflectors, optical units, lenses, semitransparent mirrors used to form reference and object beams for training neural networks, etc.

The process of photonic learning is represented in Chap. 10. The training system must have an analogue-to-digital interface, which can be created, for example, with the spin-wave holographic memory [25]. Powerful laser radiation changes the optical properties of the medium in which it propagates. Considering the effects of changes in the characteristics of the medium by sufficiently powerful waves, nonlinear transformations are needed.

Our study of the problems with creating required HM is that the requirements for this material are as follows:

- One "point" (multiple data pages) of the HM must store up to 100 quickly rewritable 3D images; for this, the material on the atomic level must ensure the saving of the same amount of different and separable quantum stable states.
- The recorded image results from the object's Fourier transform with the laser spot size less than 0.1 mm by, e.g. using a laser with a shorter wavelength and a lens with a higher numerical aperture.
- The new Fourier image is recorded by changing the inclination of the reference and object beams.
- One block of 3D rewritable holographic memory should allow matrix storage of up to five thousand images.
- The deflector system should allow discretely changing the direction of the reference and object beams.

Under these requirements, the photonic HM will ensure a better analogue than the digital one of natural human thought because it considers the behaviour of the atomic level of the human neural network, including quantum particles' non-local effects and random fluctuation of human neurons' behaviour.

5.1 Chapter Conclusion

- The meaning and purpose of human decisions are infinite. Words seem to protect the meaning of the decision—the true meaning and purpose lie in the zone of silence.
- Modern digital neural networks with billions of parameters require high energy resources and massive datasets for long-term training; human thoughts and feelings don't only have a digital and verbal nature, but also an analogue one.
- The limitation of modern digital AI tools can be overcome by applying photonic approaches; for example, light can enormously decrease computer energy consumption while increasing bandwidth and speed of signal transformation.
- Currently, computer operations of multiplication or integral transformations, as well as the learning and inference of neural networks implemented in optical systems, are based on the modulation of light passing through the transparent screen; however, they usually repeat algorithms of digital computing.
- The main problem with creating photonic AI is that a unique photonic material for 3D rewritable holographic memory has to be synthesised; laser beams must write and rewrite a finite number of images in one of the memory points; and for this, the material must be multi-stable on the group of atoms and quantum structure levels.
- It is necessary to develop an optical neural device (processor) for which unique fibres, deflectors, and rewritable holographic lenses whose light characteristics can be changed by a modulated laser beam must be made; the requirements for such a material are formulated.

References

1. Narayanan, D., Shoeybi, M., Casper, et al.: Efficient large-scale language model training on GPU clusters using Megatron-lm. In: Proceedings of the International Conference for High Performance Computing, Networking, Storage and Analysis, pp. 1–15 (2021).
2. Wu, Z., et al.: Peta-scale embedded photonics architecture for distributed deep learning applications. J. Lightwave Technol. (2023). https://doi.org/10.1109/JLT.2023.3276588
3. Rizzo, A., et al.: Integrated Kerr frequency comb-driven silicon photonic transmitter (2021). https://doi.org/10.48550/arXiv.2109.10297
4. Raikov, A.: Cognitive semantics of artificial intelligence: a new perspective. In: Topics: Computational Intelligence XVII, Springer Singapore (2021). https://doi.org/10.1007/978-981-33-6750-0
5. Kim, B.Y., Okawachi, Y., Jang, J.K., et al.: Turn-key, high-efficiency Kerr comb source. Opt. Lett. **44**(18), 4475–4478 (2019). https://doi.org/10.48550/arXiv.1907.07164

6. Wang, W., Khazraee, M., Zhong, Z., et al.: Topoopt: Co-optimizing network topology and parallelization strategy for distributed training jobs (2022). https://doi.org/10.48550/arXiv.2202.00433
7. Lu, Y., Gu, H., Yu, X., Li, P.: X-nest: a scalable, flexible, and high-performance network architecture for distributed machine learning. J. Lightwave Technol. **39**(13), 4247–4254 (2021)
8. Chen, Y., Nazhamaiti, M., Xu, H., et al.: All-analog photoelectronic chip for high-speed vision tasks. Nature **623**, 48–57 (2023). https://doi.org/10.1038/s41586-023-06558-8
9. Bandyopadhyay, S., et al.: Single-chip photonic deep neural network with accelerated training (2022). https://doi.org/10.48550/arXiv.2208.01623
10. Gyger, S., et al.: Reconfigurable photonics with on-chip single-photon detectors. Nat. Commun. **12**, 1408 (2021). https://doi.org/10.1038/s41467-021-21624-3
11. Novack, A., et al.: Germanium photodetector with 60 GHz bandwidth using inductive gain peaking. Opt. Express **21**, 28387 (2013)
12. Gamboa, J., et al.: Ultrafast image retrieval from a holographic memory disc for high-speed operation of a shift, scale, and rotation invariant target recognition system (2022). https://arxiv.org/ftp/arxiv/papers/2211/2211.03881.pdf
13. Jordan, S.P., Krovi, H., Lee, K.S.M., Preskill, J.: BQP-completeness of scattering in scalar quantum field theory. Quantum (2018). https://doi.org/10.22331/q-2018-01-08-44
14. Nishioka, T., Ryu, S., Takayanagi, T.: Holographic entanglement entropy: an overview. J. Phys. A **42**, 504008 (2009). https://doi.org/10.48550/arXiv.0905.0932
15. Susskind, L.: Three lectures on complexity and black holes (2018). https://doi.org/10.48550/arXiv.1810.11563
16. Ali-Akbari, M., Lezgi, M.: Resource and stability near a critical point from the quantum information perspective (2022). Phys. Lett. B **842**, 137954 (2023). https://doi.org/10.48550/arXiv.2209.04623
17. Lightelligence. https://www.lightelligence.ai/. Accessed 30 Nov 2023
18. Sainath, T.N., Kingsbury, B., Saon, G., Soltau, H., Mohamed, A.R., Dahl, G., Ramabhadran, B.: Deep convolutional neural networks for large-scale speech tasks. Neural Netw. **64**, 39–48 (2015)
19. Wu, C., Yu, H., Lee, S., Peng, R., Takeuchi, I., Li, M.: Programmable phase-change metasurfaces on waveguides for multimode photonic convolutional neural network. Nat. Commun. **12**(1), 1–8 (2021)
20. Ashtiani, F., Geers, A.J., Aflatouni, F., Ashtiani, F., Geers, A.J., Aflatouni, F.: An on-chip photonic deep neural network for image classification. Nature **606**, 501–506 (2022). https://doi.org/10.1038/s41586-022-04714-0
21. Xu, X., et al.: Photonic perceptron based on a kerr microcomb for high-speed, scalable, optical neural networks. Laser Photonics Rev. **14**(10), 2000070 (2020). https://doi.org/10.1002/lpor.202000070
22. Hamerly, R., Bernstein, L., Sludds, A., Soljačić, M., Englund, D.: Large-scale optical neural networks based on photoelectric multiplication. Phys. Rev. X9, 021032 (2019). https://doi.org/10.48550/arXiv.1812.07614
23. Brambilla, N., Vladyslav Shtabovenko, V., Castellà, J.T., Vairo, A.: Effective field theories for van der Waals interactions (2017). https://doi.org/10.48550/arXiv.1704.03476
24. Fischer, C.F.: General Hartree-Fock program. Comput. Phys. Commun. **43**(3), 355–365 (1987). https://doi.org/10.1016/0010-4655(87)90053-1
25. Gertz, F., Kozhevnikov, A.V., Filimonov, Y.A., Nikonov, D.E., Khitun, A.: Magnonic holographic memory: from proposal to device. IEEE J. Explor. Solid-State Comput. Dev. Circuits **1**, 67–75 (2015). https://doi.org/10.1109/jxcdc.2015.2461618

Chapter 6
Soliton Thoughts

Laser solitons can help to construct photonic AI. They are waves with stable structures, and their combinations resemble human thoughts' overlays. They move in space and combine complexes without distorting each other. The coupling of solitons remains weak throughout evolution while all the closed energy lines of every soliton are preserved. Semiconductor interferometers, fibre light guides, and lasers are used for the optical processing of solitons.

6.1 Laser Solitons Recall Human Thoughts

Human thoughts can be intrusive and stable, and they are also elusive and unstable. Unwritten thoughts quickly disappear, and recorded ones undergo semantic distortions over time. Solitons are generated and formed using a laser and an optical coherent wave modulator in a wave synthesiser.

Solitons exist while the laser is working, and the conditions required to maintain them are preserved. Therefore, storing every soliton with its semantic interpretation by a hard optical (long-term) memory is necessary to build artificial analogies of mental processes using solitons. For example, such a stored semantic interpretation can be made by holographic tools.

Weak radiation in a homogeneous medium is diffused in the transverse directions due to diffraction, and in the longitudinal direction due to the dispersion of the medium, that is, due to the difference in the effect of the properties of the medium on radiation with different wavelengths. Symmetric solitons can be used to construct asymmetric moving connected complexes of solitons. It is possible to organise collisions of such complexes. Collision scenarios can be very diverse and include variants with a change in the number of solitons.

Solitons are of two main types: conservative and dissipative [1]. The former exists in optical systems with weak dissipation, allowing for a relatively long lifetime of the

A. Raikov, *Photonic Artificial Intelligence*,
SpringerBriefs in Computational Intelligence,
https://doi.org/10.1007/978-981-97-1291-5_6

soliton, perhaps even infinite. The latter arises due to the inflow and outflow of energy balanced in a certain way. Dissipative solitons are stable localised field structures in a homogeneous or weakly modulated nonlinear medium. Their lifetime depends on the balance of energy in these flows. A human thought is better associated with dissipative solitons.

Light rays in a medium with an inhomogeneous refractive index bend towards a medium with a higher refractive index. In a laser beam of light, the intensity is maximal on its axis and decreases to the periphery. Then, if the refractive index of the medium increases with increasing light intensity, the beams will bend to the laser beam axis, and the medium becomes equivalent to a focusing lens. Such focusing compensates for linear diffraction blurring, contributing to the generation of a conservative soliton.

Dissipative optical solitons can be stationary, moving and rotating, static and changing periodically or chaotically, single and connected. For example, the wavefront of vortex solitons includes dislocation, i.e. the radiation intensity at a certain point turns to zero, and the phase, when completely bypassing this point along a closed contour, shifts by a certain amount called the topological charge. There is a vortex motion of radiation energy around the centre of such solitons.

Dissipative laser solitons can be one-dimensional, two-dimensional, and three-dimensional. Thus, in two-dimensional solitons, the longitudinal change in the amplitude of the field is small, the nonlinearity of the medium is not inertial, and the polarisation of the radiation is close to linear. Therefore, instead of the phase, it is advisable to consider the transverse energy flux of radiation, which is determined at a slight angular radiation divergence by the product of intensity by the transverse gradient of the phase of the field.

The portrait of a dissipative soliton can be associated with symbols. The centre of the soliton can be a fixed point of energy flows. Other elements are closed curves, such as circles dividing the portrait into cells. In the case of vortex dissipative solitons, the vortex flows resemble a whirlpool, and there may be more cells in them. The circles are the limit cycles of other flows. Thus, dissipative solitons have an internal structure determined by the topology of energy flows.

Suppose several dissipative solitons parallel to the laser axes are excited in a wide-aperture laser. In that case, the interaction between them will be determined by the degree of overlap of their fields. With a small degree of overlap of solitons, the interaction is weak and depends on the distances and phase differences between them. The connection between solitons is considered weak if all the closed lines before the overlap are preserved in the portrait of energy flows. In this case, it is possible to construct quantum-like mechanics of solitons, treating them as particles.

Dissipative solitons behave like a quantum particle: there is a discrete set of their states (characteristics); they are excited in a threshold manner; for the appearance of a dissipative soliton, a sufficiently large initial energy release is needed—above a certain threshold. If the amplitude of the ejection is less than the critical value, then the soliton resolves. In the coupled pairs, the solitons behave in phase or antiphase, while there is a discrete set of equilibrium distances between them. Stable asymmetric structures can be constructed from many identical dissipative solitons. For example, structures in an isosceles triangle are established from a trio of solitons with phase differences between pairs.

A pair of parallel flat mirrors form the laser resonator, the energy source is energy pumping, and energy drains occur through a specially placed absorber, modulator, and other loss channels in the laser. The laser parameters are selected so that there is no radiation generation if the losses exceed the gain at low radiation intensity. With an increase in radiation intensity, the gain increases faster than the absorption; therefore, at a certain level of intensity, their stable balance and support for radiation generation are possible. The laser can be with a large cross section—the actual widths of solitons in semiconductor resonators are about a dozen micrometres. Then, a generation mode can be established on the central part of the aperture, and a non-generation mode can be established on the rest. Diffraction blurs the transition between modes, and then the corresponding bright spot on the dark background of the non-generating mode will represent the simplest laser dissipative soliton.

Dispersion solitons have an analogue of quantum properties—discrete rather than continuous states. At the same time, such a soliton has an internal structure associated with the scheme of energy flows. A dissipative soliton can be compared with a discrete atom's energy levels. When several such "atoms" interact, "molecules" are formed, and zones of states are formed in the lattice of "atoms" as in a crystal. Laser solitons as "atoms" can be used to build both "ideal" (without defects) and "real" (with defects) crystals that move, including curvilinearly, and rotate with a set of linear and angular velocities. With an increase in the number of coupled solitons, their spectrum thickens and becomes continuous for an infinite chain of solitons. This soliton modelling of the thinking process also suggests that thinking does not follow classical modelling approaches.

As already noted, dissipative solitons are different—there are types: stationary and pulsating; moving with constant and variable speed, rotating; with a regular wavefront and dislocations of higher orders; scalar and vector, unique- and multi-frequency; single and coupled; one-dimensional, two-dimensional and three-dimensional; characterised by continuous and discrete spectra.

Thus, the variety of forms of optical dissipative solitons combined with the possibility of controlling them using lasers suggests the opportunity of their use to represent "symbols of thought" in an optical medium, which itself has quantum mechanical and relativistic features. This can describe weakly formalised cognitive semantics [2] of advanced (general or strong) AI models built in the alphabet's language formed with the help of solitons. Such a soliton-alphabet can have, for example, an image of hieroglyphs. However, to represent the cognitive semantics of symbols and models constructed from them:

(a) The shape of a soliton must be generated by artificial modulation of a laser beam (wave),

(b) Each "soliton symbol" must be mapped to a relevant array of big data from optical images of objects that fill this symbol with meaning.

The theory of solitons is currently implemented in practice, mainly in information transmission via communication channels. For example, in fibre light guides, an impulse with a shape described using a hyperbolic secant propagates relatively long distances without distortion. Such a transmitting device may consist of a laser, an insulator, a modulator, a fibre-optic channel, and a receiver. In addition, amplifiers can be supplied throughout the transmission line. Such soliton systems in distributed AI systems can help transmit signals between agents at a great distance.

Building an Artificial Mind requires a unique system for representing the outside world (see Chap. 2). This can be a laser resonance system in which dissipative solitons of a given modulator shape are generated. The number of solitons' forms should be limited. Still, this number should be enough to identify several million concepts that reflect and cover relevant areas of knowledge to the research topic. Each identifier of the concepts, a soliton, or a combination of solitons, should be reflected on the maximum possible knowledge in images of articles, books, reference books, etc.

6.2 Laser-Soliton Equipment

A person needs a reason to achieve the desired results purposefully. The results themselves may be known in advance. For example, the task is to determine the history of the Universe's dynamic during the Plank epoch (it refers to the earliest period in the Universe's history, from zero to 10^{-43} s after the hypothetical Big Bang). The theory of quantum gravity is needed to understand this period better. Cosmic string theory can help answer this issue; however, no experimental confirmation exists [3].

The mind often solves an incorrect problem: the goal is known inaccurately, the existence of a solution is doubtful, there may be several solutions, the resource is limited, a path converging to the goal still needs to be found, and there may be many such paths. Therefore, mind-solving tasks are seen as the inverse problem-solving since their solution comes from a goal still unknown but anticipated in the human mind in general terms. However, by definition, any problem (task) requires formulation regarding some symbolism. Such a set of symbols can be, for example, printed language. But the mind operates not only with language but also with unprintable meanings, for the description of which the AI tools currently used, such as logical ontologies and neural networks, need significant replenishment with quantum wave, photonic, and other means to represent cognitive semantics and hidden meanings [2].

Thought has a certain autonomy and does not have a clear linguistic representation. To paraphrase Hegel, a symbol (predicates, language) "spirally" denies thought, although language, identifying thoughts, makes them communicable with other thoughts in one person and between different people. As the thought has, at least, a biological-quantum-wave character, it is influenced by feelings and external pressure, which feeds the thought through biological perceptual and nervous channels, which, in turn, may have a non-local nature of the behaviour, that is, be influenced by the near and far external environment.

Considering the soliton and biological-quantum-wave interpretation of the thought process, they can be represented using an optical medium. The idea of creating a corresponding soliton-based system is to replace a printed symbol or a symbol represented digitally in a semiconductor chip with its quantum-wave analogue in the form of a soliton. Such a replacement simultaneously compensates to some extent for the deficiency of a biological component of the thinking mechanism in digital computers. At the same time, each symbol-soliton must be matched with its materialised semantic interpretation in the complex form, for example, by 3D holographic images of a relevant set of information objects. The latter can serve as the optical representations of descriptions of thoughts, dreams, feelings, emotions, mind images, sounds, and voices—with their quantum-wave peculiarities.

The coupling of the dispersion soliton with its semantic interpretation in holographic memory may be controlled discretely through a digital control system, including a supercomputer. To do this, the modulator has to generate a given soliton synchronised with the allocating of a set of Fourier images in a holographic storage device corresponding to its semantic interpretation. During the lifetime of each generated soliton in the wave synthesiser, the selection of its semantic interpretations remains constant. Then, combining two or more solitons will create the union of sets of optical semantic interpretations of these solitons.

Feelings and emotions fulfil the cognitive semantics of AI models [2]. The photonic semantics approach is continuously compared to the formalised cognitive semantics of traditional digital AI models. This approach helps recognise and encode feelings and emotions by faces, texts, voices, and encephalograms in an optical-quantum-wave manner (see Chap. 3). To do this, each position (dot) in a continuous many-dimensional space stored in holography memory of emotions must be mapped to a math convolution of different images of one of the emotions (see Chap. 10).

An optical soliton is an impulse in the form of a single bell-shaped wave formed in an optical system, for example, a fibre, in the presence of a nonlinear dependence of the refractive index on the radiation intensity of a coherent source. At the same time, the refractive index increases with increasing intensity, and the high-frequency components of the soliton shift to its tail and the low-frequency ones—to its head, suppressing chromatic, and polarisation dispersion. Such a pulse can retain its shape along the entire length of the optical system. Solitons with specific support can propagate in optical fibre for several thousand kilometres with almost no shape distortion and persist in collisions with each other.

The uniqueness of the soliton is that a nonlinear change in the refractive index completely balances the dispersion of the soliton group velocity. Therefore, a fairly accurate description of the conditions for the existence of optical solitons is obtained by solving the Schrodinger equation [4, 5]. The conditions for the existence of solitons have been thoroughly evaluated in various media and considering the signal power, the radius of the mode spot in the glass fibre, the central frequency of the signal spectrum, the refractive index of the fibre core, pulse duration, dispersion coefficient, the collision period of solitons, amplification in a fibre amplifier, etc.

Solitons break up into groups and then gather again, which looks like the behaviour of the thought process. Factors such as phase self-regulation, dispersion of group velocities, power, pulse duration, etc., determine these dynamics. In a soliton laser, very short pulses (units of picoseconds and femtoseconds) are generated, which can be achieved by compressing pulses with a duration of nanoseconds.

The soliton model can first be constructed in a classical way using Fourier optics methods and stochastic nonlinear partial differential equations. Then, the numerical solutions of the stochastic models of solitons have been estimated and simulated [6–8]. Finally, the soliton model can form the corresponding optical image displayed in the modulator.

A modulator is a device for a specific change in the intensity of a transmitted laser coherent beam using its diffraction on a lattice formed in glass due to spatial modulation of the refractive index. To do this, the optical system must have a corresponding block. The modulator can be built using a multidimensional holographic storage device, in which many—up to several thousand and more—soliton patterns are recorded. Then, the necessary soliton can be illuminated by the direction of the reference laser beam at a certain angle to the appropriate place of the holographic storage.

The soliton transformation takes place in a wave synthesiser. A laser and a corresponding beam expander generate a coherent light beam with a wide aperture. The soliton synthesiser can be built based on a sufficiently long optical wave cable. Appropriate signal amplifiers and sensors can be installed along the cable's length to detect the result of the generation and combination of solitons.

The laser soliton system may be used for inverse problem-solving. For this, the target soliton must be previously created. The combinations of solitons diagnosed by sensors are evaluated for compliance with the target soliton; the result is fed to the control device, and, as a result, the next set of solitons is modulated. This can be done in the order specified by the genetic algorithm. New solitons may be synthesised by combining laser soliton and its interpretation's optical semantics in holographic memory. The newly synthesised soliton has to be compared with the target soliton, for which the corresponding Fourier convolution may also be done in holographic memory. To solve the inverse problem of constructing a path to the target soliton, the generation of solitons must be made many times.

The optical genetic algorithm works step-by-step by creating the combined laser solitons, the shape of which consistently converges to the shape of the target soliton. The parameter of the proximity of the soliton forms or some quality function can be introduced, for example, using the optical convolution operation of the compared solitons. To implement one, it is necessary to ensure the execution of four operations with the connection of the mechanism of random number generation:

(a) Generation through modulation of the initial population of solitons.
(b) Reproduction (selection)—construction of a new population of solitons through a random selection from the available ones, in which the value of the quality function is higher than the original one, and the proximity to the target soliton will be higher.

(c) Crossover—construction of a new population of solitons in which new solitons will appear due to a pairwise combination of fragments of solitons, in which the proximity to the target soliton will be higher.

(d) Mutations—constructing a new population of solitons with a better proximity characteristic by replacing one of the solitons of the generating population with a new one.

Optical soliton technologies can realise information storage and transmission, but external signals can modulate solitons. This creates certain limitations in practicability. For example, the laser beam may be split into many beams, but the quantum non-locality effect can prevent the catastrophic collapse of high-dimensional beams and suppress modulation instability. As a result, a range of solitons is supported due to the long-ranged interaction of solitons. Therefore, it is still challenging to design an optical soliton device with stable soliton states and more convenient operation based on laser [9].

The building of cognitive semantics of solitons by mapping them to sets of holographic images is accompanied by quantum effects: decoherence, quantum correlation (the entanglement), a change in the state of quantum particles, wave function collapse, etc. These effects must be considered when creating a photonic AI system. Thus, the states of the quantum particles that the solitons consist of should be described using a density matrix. This matrix contains a probability distribution for possible measurement options.

Considering quantum effects forces us to think about the behaviour of the quantum particles that make up the solitons, in the form of, for example, particles and waves or particles that are accompanied by "shadow" particles. This raises the issue of fulfilling the quantum problematic characteristics of thoughts, see Chap. 7.

6.3 Chapter Conclusion

- Laser solitons, which are electromagnetic waves that move and rotate, changing periodically or chaotically, may be single and connected and emerge non-local quantum effects—look like human thoughts, which can be intrusive, elusive, unstable, autonomos, and do not have a clear linguistic (digital) semantics interpretation.
- Solitons recall quantum particles: there is a discrete set of their states; they are excited in a threshold manner—an initial energy release is needed for their appearance.
- Solitons are different: stationary and pulsating, moving with constant or variable speed, with a regular wavefront and dislocations of higher orders; unique- and multi-frequency; single and coupled; one-, two- and three-dimensional; characterised by continuous and discrete spectra.
- Soliton transformation occurs in a wave synthesiser, which can be based on an optical wave cable of a sufficiently long length.

- The inverse problem may be solved with a genetic algorithm using laser solitons combined with their optical (holographic) semantic interpretation.
- The idea of creating a photonic AI system for making non-symbolic thought operations may be realised based on laser solitons and holographic memory, representing cognitive semantic interpretations of solitons-thoughts.

References

1. Rosanov, N.N.: Spatial hysteresis and optical patterns. Springer, Berlin, Heidelberg (2013). https://doi.org/10.1007/978-3-662-04792-7
2. Raikov, A.: Cognitive semantics of artificial intelligence: a new perspective. In: Topics: Computational Intelligence XVII. Springer, Singapore, p. 128 (2021). https://doi.org/10.1007/978-981-33-6750-0
3. Sazhina, O.S., Scognamiglio, D., Sazhin, M.V., Capaccioli, M.: Optical analysis of a CMB cosmic string candidate. Mon. Not. R. Astron. Soc. **485**(2), 1876–1885 (2019)
4. Zhu, C., Abdallah, S.A.O., Rezapour, S., Shateyi, S.: On new diverse variety analytical optical soliton solutions to the perturbed nonlinear Schrödinger equation. Results Phys. **54**, 107046 (2023). https://doi.org/10.1016/j.rinp.2023.107046
5. Guo, M., Xie, X.: Binary Darboux transformation and interactions of solitons for a higher-order matrix nonlinear Schrödinger equation. Results Phys. **53**, 106942 (2023). https://doi.org/10.1016/j.rinp.2023.106942
6. Baber, M.Z., Ahmed, N., Yasin, M.W., et al.: Comparative analysis of numerical with optical soliton solutions of stochastic Gross-Pitaevskii equation in dispersive media. Results Phys. **44**, 106175 (2023). https://doi.org/10.1016/j.rinp.2022.106175
7. Yasin, M.W., Iqbal, M.S., Ahmed, N., et al.: Numerical scheme and stability analysis of stochastic Fitzhugh-Nagumo model. Results Phys. **32**, 105023 (2022). https://doi.org/10.1016/j.rinp.2021.105023
8. Yasin, M.W., Iqbal, M.S., Seadawy, A.R., et al.: Numerical scheme and analytical solutions to the stochastic nonlinear advection-diffusion dynamical model. Int. J. Nonlinear Sci. Numer. Simul. (2021). https://doi.org/10.1515/ijnsns-2021-0113
9. Han, H., Wang, R., Cao, H., et al.: Coded information storage pulsed laser based on vector period-doubled pulsating solitons. Optics Laser Technol. **158**(A), 108894 (2023). https://doi.org/10.1016/j.optlastec.2022.108894

Chapter 7
Subatomic Thinking

Human neurons are made up of atoms, and their behaviour and thinking processes depend on the dynamic of matter on the subatomic level. Immersing into the atomic level of representing the thinking processes by the AI models should consider the quantum effects, including non-causal and non-local ones.

Imagine standing on a warm early morning by the sea beach and watching the sunrise. The sea air, the smell of flowers, the wet sand, and the sound of rolling waves please your imagination and fill you with new impressions. At the same time, you are immersed in memories, thinking about life issues and arousing future dreams (Fig. 7.1).

Is it possible to replace this emotional state of feeling with a digital AI system, including neural and generative? It is hardly likely. Diving significantly more profoundly into the structure of neuron matter is necessary than its digital and logical representation in modern computers.

Matter consists of molecules; molecules are made of atoms, and atoms are made of elementary particles; among many elementary particles embraced by the Standard Model are leptons, quarks, fundamental bosons, and so on [1]. Fields are associated with matter and particles: electromagnetic and gravitational, strong and weak. An atom consists of a nucleus and electrons. Nuclei consist of protons and neutrons, interconnected by gluons—quanta of strong interaction, strong force, or strong nuclear force. Such a representation of human neurons requires an infinite deepening into the structure of matter, including quantum particles, quarks, and photons.

Protons and neutrons are made up of quarks. Quarks cannot be single; they are part of particles with a whole electric charge. Quarks and electrons are truly elementary particles. At the modern level of knowledge, they do not consist of anything; however, they can turn into other particles. These elementary particles and a photon are enough to build almost all the matter on Earth, including the human brain.

These particles are from the first generation of elementary fermions. There are also second and third generations of particles that have a larger mass. The fermion corresponds to its antiparticle. Particles interact, realised by exchanging particles

Fig. 7.1 Watching the sunrise (the author made the photo)

with field quanta by some virtual particles. Some interactions are noticeable only at very small distances, such as weak interaction. It is more correct to look at these interactions not as forces but as transformations.

Each field corresponds to a particle, and each particle corresponds to a field. At the current level of understanding of physics, field quanta are entirely devoid of individuality; that is, it is impossible to number particles of the same kind and track the movement of each of them. So far, it is believed that the photon is not a composite particle; it does not divide into smaller parts. *Then, why is the photon fluctuating? Does something make him do it from the inside?*

It is worth dwelling on this unstructured property of the indivisibility of the photon in more detail. According to the current interpretations of quantum physics, the flight of a single photon is accompanied by an infinite number of shadow photons or waves. This effect can be non-direct detected in the "double-slit" experiment when a stream of single photons interferes with shadow photons or waves and forms an interference spectrum on the screen. However, these shadow sides of photons cannot be detected because observation causes collapse—the interference pattern of the photon disappears.

These properties give rise to the hypothetical idea that a single photon itself carries information about the source, for example, a star from deep space, a laser, etc. Perhaps this information is not distorted by the aberrations of optical devices receiving and transforming light rays containing photons. To prove the correctness of this idea, we can formulate a hypothesis about the possibility of recognising the spectrum of a single photon radiation source based on the double-slit effect. A device with the following basic units is required to confirm the hypothesis: a telescope, a photon filter that samples single photons from a continuous light beam, and a corresponding double-slit system. The telescope receives a signal from a distant star, the filter transmits a stream of single photons, and the double-slit forms an interference spectrum.

The experiment may be carried out on several sources: two stars, one star and a laser, and two lasers. The goal is to prove that a single photon can carry information about the source, generating its unique interference spectrum and having a unique structure of behaviour. A unique interference pattern, if such a thing happens, through inverse problem-solving allows us to determine the spectrum and, accordingly, the characteristics of a distant source. The difference in the spectra of two stars obtained using the proposed device will confirm the correctness of the appropriate interpretation of quantum physics with the presence of shadow photons.

It is believed that shadow photons (if they exist) do not carry information about the source due, for example, to the fact that the spectrum of the wave packet of a single photon is limited and that these shadow photons cannot be detected. The positive result of the proposed experiment, in addition to confirming the appropriate interpretation of quantum physics, will also be helpful for practical use:

- There will be no need to develop costly optical, including space, telescopes.
- The understanding of the photon characteristics will be deepened.

The detected fact of photon fluctuation may indicate the presence of a photon's ability to carry unique information about the source. The behaviour of photons is more complex than described in classical physics. The paper [2] shows an efficient computational image recognition algorithm to fuse the spatial frequencies from the classical low-spatial-frequency range with the spatial frequency content in the quantum signals. For this, multiple orders of the Glauber correlation function were extracted. As known from classical optics theory, spatial frequencies higher than $1/\lambda$ (λ is a wavelength) decay exponentially in amplitude during propagating from a recognised object. However, the unique quantum properties of photons can be exploited to get additional information by detecting quantum correlations [3].

Detection of fluctuations in light beam's photons emitted by the recognised object can make it [4, 5]. Correlations in the emitted light can be used for scalable enhancements in imaging resolution. Classical correlations are times greater than microseconds, and quantum correlations occur in submicrosecond periods [6]. The photon fluctuations depend on the emitters' characteristics, the environment, the integration time, and the detector tool. The statistics of classical temporal photon fluctuations can be used for computational super-resolution. New information can be obtained by calculating correlations between the detected signals from fluctuating photons.

The idea is as follows: a single emitter can only emit one photon at a time because the emission only occurs when the emitter is in an excited state. Correlations of photons from a single emitter have a null at zero delay between photons, which emerged due to applying one code to the wave packet. This property is called anti-bunching. It can be used to distinguish between the spatial locations of quantum emitters.

Particles can have internal degrees of freedom. Such degrees can characterise charges and the intrinsic moment of momentum—spin. The state of a particle's internal degrees of freedom can also be described by polarisation. The angular momentum of the particles can be measured in Planck constant. The moment associated with the motion of a particle is always equal to an integer, and the spin can be either an integer or a multiple of 1/2. Particles with half-integer spin are fermions. There can be no more than one fermion in each state. Particles with a whole spin are bosons. They are "friendly" and strive to be in the same state simultaneously.

There are more than ten interpretations of quantum mechanics, so the question arises: to what extent and what interpretations dictate universal principles that can be applied to human thought activity? So, in our work [7], within the framework of the particle–wave interpretation of quantum physics, an attempt was made to identify the analogy of the behaviour of an elementary particle and a word in human speech and thoughts. The word itself is likened to a particle, and the cognitive semantics of the word are likened to a wave. At the same time, for the construction of cognitive semantics of AI models, the semantic aspects of the representation and interaction of particles are of interest, not only the physical aspects that underlie the relationship of particles.

Such an association around the cognitive semantics of AI models suggests the explanation of the process of people's communication. Two people are talking; each of them knows himself better than the other and does not know what is going on in the feelings and thoughts of the interlocutor. Although they know about emotions, decisions, and thoughts inaccurately, at any moment, they can change: sad feelings turn into joyful ones, pain in the head quickly passes, etc. Here, we see an association of thinking processes with a single photon flying towards the diaphragm and the screen behind it, which "does not know" where it will fall on the screen, flying through the aperture.

By this association, quantum particles can also be endowed with semantic features. After all, a quantum particle is, in fact, an abstraction. For example, in the Copenhagen interpretation of quantum mechanics, only the results of actual observations and measurements are stated. This interpretation does not deny the real world but notes the fundamental impossibility of a detailed analysis of the interaction in the natural world and not the abstract in the form of a formula, a quantum object, and a measuring device.

In another interpretation of quantum mechanics, it is proposed to consider the existence of the Universe as the result of the observer's participation, that is, his immersion in reality: there is no observer–no object. There are many variants of the observation result, and the wave function reduces to one of the possible variants precisely at the moment of observation, measurement: the completeness of the phenomenon appears after the observer takes a measuring device (for example, a telescope) and directs it to the object.

There is also a contrary opinion, insisting on excluding the "subjective element associated with the observer" from quantum mechanics. The primary role is given to the "arrow of time". The phenomena are irreversible, and reversible processes are just an exception. At the same time, on a different scale, with modern consideration of the Big Bang model and the expansion of the Universe, it is not yet possible to say what prospect awaits it in the distant future. Maybe it will "shrink" because, for example, the idea of an infinitely fractal Universe can be falsified [8].

From the point of view of choosing the principles of constructing cognitive semantics [7], the known interpretation of quantum physics in the form of a plurality of worlds is clearly of interest [9]. Such an interpretation can be fruitful in AI models' macrocosm, non-local semantics, communications, and discourses.

Quantum interpretation of von Neumann, Birkhoff, etc., is based on the non-classical logical paradigm, in which a fetish is formed regarding the ontological self-worth and content status of logics, that is, linguistics, and their attribution to the world, and not to language.

The holistic interpretation of quantum mechanics works on the integrity of the representation of the semantics of AI. The Universe is a hologram, and each of its points reflects and stores data on the Universe. The objects themselves, actions, and movements are objects of the hologram. This approach claims the status of establishing universal order in the Universe since holographic transformations imply a particular order, instruction, and physical constitution.

From the point of view of constructing cognitive semantics [7], the interpretation of quantum mechanics can introduce the concept of a quantum ensemble, which resembles the concept of an ensemble from statistical thermodynamics. A quantum ensemble is formed by the endless repetition of a situation created by the same microsystem immersed in the same environment, which, in turn, sets the microsystem in a particular "state". This interpretation pays less attention to the observer and emphasises quantum ensembles' objective nature.

The interpretation of quantum mechanics proposed by R. Feynman is still relevant. In contrast to relying on the Schrodinger equation, it is suggested to take an integral from the Lagrangian, which allows representing a time-dependent wave function through its initial value. From the point of view of organisational structuring, it is essential to note that such a procedure will enable the creation of both classical trajectories and all *imaginable trajectories* connecting the starting point and the point at another point in time.

It is also necessary to highlight the interpretation proposed by Heisenberg and Fock. Unlike the Copenhagen interpretation, their interpretation asserts that there is no reality behind the quantum phenomenon but in a different sense. Karl Popper suggests considering the wave function as describing additional properties to the classical one, as predispositions of the behaviour of objects. Latent and manifest aspects of possible phenomena are highlighted in the work [10]. In most interpretations, the subjects do not participate in the measurement; they use the metre to evaluate the results.

Thus, focusing on the observer and semantic aspects of quantum representations of natural human neurons, these aspects can be reduced to the following, as yet incomplete, list [11, 12]:

- The behaviour of quantum particles is described in an infinite-dimensional space, and each particle's state is represented through only six characteristics, including mass and momentum; such reductionism is insufficient to artificially interpret human consciousness and mind by classical digital AI systems.
- In quantum interpretations, a holistic view of the Universe stands out; still, such a view is paradoxical since, on the one hand, everything is connected, and on the other, for successful thinking, it needs to be given separability, objectivity, and focus.
- For a quantum particle to be in a "different place" means to be another particle; this is well associated with a change in the semantic interpretation of the AI model when the environment or situation changes.
- A quantum particle and a wave behaviour are well associated with the nature of human discourse, depending on the context and goals of the discourse's participants.
- Any attempt to detect a wave accompanying a particle collapses the double-slit interference effect; this is well associated with attempts to verbally represent thoughts—the word reduces thought content.
- The change in the state of a quantum particle occurs abruptly, and the reason for such a change is unknown; the particle is forbidden to have zero values of combinations of its parameters (for example, location and momentum); this feature may resemble the effect of enlightenment in wisdom persons.
- It is impossible to measure a particle state autonomously since the effect of non-locality is characteristic of a particle, that is, the dependence of its behaviour on another particle, the position, and momentum of which are not precisely defined; this suggests the ancient idea that human thinking is conceptually part of some hypothetical universal consciousness.
- The quantum approach to AI modelling, considering non-local semantics of particles, tries to delve into the phenomena themselves and thereby ensure the fusion of the object under study with the model, turning the simulated object into the actual cognitive semantics of the AI model.

For about a hundred years, there has been a dispute about the structure of reality and the finality of describing it in a quantum mechanical way. For example, the aspect of the non-locality of quantum phenomena encourages the activation of a long-standing discussion about the universality of the causality principle. For instance, at the quantum level, actual events demonstrate the possibility of causeless events.

On this topic, it is necessary to highlight the well-known entanglement effect [13] and the Bell test for local realism [14]. Apparently, Bell's works of 1964–1966 became the final turning point when the construction of physical theories ceased to be carried out within the conceptual framework of classical physics. It remained only to confirm this experimentally, but there was no such technical possibility then.

And so, thanks to the development of information and telecommunication technologies, the possibility of complete unpredictability of phenomena, which means non-causality, manifestations of human "free will", has recently been confirmed [15]. For this purpose, about 100,000 people participated in a specially organised experiment simultaneously. Such a result sets the stage for constructivist approaches that consider subjects' extraordinary creative and active role in real life, creating the basis for a decisive verification of the existing pictures of reality, including those that differ from representations of quantum mechanics.

It is worth noting Bell's work in the field of hidden parameters research. Since quantum mechanics is described by probabilistic theory, the description of the state of the system cannot be accurately represented; it is possible to calculate and predict quantum events only taking into account some probability. That is, the description of the system's state using the wave function is incomplete, and perhaps some "hidden" parameters are not available for observation using currently available tools and methodological approaches. This is somewhat reminiscent of the hypothetical existence of "shadow" quantum particles that interfere with fundamental particles or the hypothetical component of matter in the form of "dark" matter, which manifests itself only through gravitational effects and does not absorb, reflect, or emit light.

The listed features of quantum semantic interpretations of AI give rise to the hope that in future, it will be possible to bridge the principles of constructing semantics of AI by using various scales and qualities, including between denotative (digital) and cognitive (non-formalisable) semantics, between objects of inanimate and subjects of living nature [7].

Thus, the non-causal and non-local quantum effects come into force if a neuron's atomic structure is considered when building a perspective AI, particularly a strong AI [16]. At the same time, it is necessary to consider the very nature of the atomic structure of matter and human neurons, which will remain a mystery for a long time. To do this, it is enough to look at the features of the structure and behaviour of atoms.

Atoms can absorb energy and emit electromagnetic waves. The emission and absorption spectra consist of spectral lines. For example, the spectrum of atomic hydrogen in the visible part of the spectrum contains four (some show five lines) spectral lines. Lines are also present in the invisible parts of the spectrum: infrared and ultraviolet. In the latter, as the frequency increases, the intensity of the lines and the distance between adjacent lines decrease.

The structure of the atom model has been presented in different ways. For example, in the planetary model of an atom, an electron moves uniformly around a circle at a certain speed. To describe the behaviour of atoms, postulates are introduced, such as the state of an atom with minimal energy is called the ground state, and all the others are called excited states; a relationship is set between the values of the energy of the atom and the frequencies of the emitted radiation, as a result of which one photon of a specific frequency is emitted.

However, the existing atom models have disadvantages due to the internal inconsistency of using the principles of classical physics with quantum postulates. For example, quantum transitions may not always explain the intensity of spectral lines or the behaviour of an electron in an atom when exposed to an external magnetic field, and, importantly, they are unsuitable for multi-electron atoms.

In multi-electron atoms, electrons move in the nucleus field, which differs from the Coulomb field due to the shielding action of electrons closer to the nucleus. The electric field of a complex atom decreases with distance from the nucleus faster than the Coulomb field of a hydrogen-like atom since the energy of an electron in stationary states will depend on the main and orbital quantum numbers.

The solution of the quantum mechanical problem for such an atom implies finding a multi-electron function describing the states of all electrons and the energy of the stationary states of the atom by solving the stationary Schrodinger equation. This equation has no strict analytical solution, and approximate methods are usually used. It is also worth noting that this equation in modelling has uncertainty limitations; for example, in the absence of external influence, it cannot predict in advance the electron jumps from one level to another; uncontrolled internal fluctuations influence this.

Speaking about the regularity of filing shells and energy levels of atoms with electrons, according to the classical paradigm, all electrons should be in a state with minimal energy. But this is not the case; the electrons encounter a restriction called the "organisational" *Pauli prohibition principle.* Its essence is that two fermions cannot simultaneously be in the same quantum state; only two electrons can be in a quantum state characterised by three quantum numbers. This principle does not apply to "friendly" bosons—particles with a whole spin. The photon belongs to the set of boson particles.

The order in which electrons fill their energy levels is determined using Hund's rule: the state of the atom with the most significant spin quantum number has the lowest energy. The regulations of Madelung and Klechkovsky determine the order of filling the electron layer with electrons: the energy of an atom increases sequentially as the sum of the primary and orbital quantum numbers increases, with the same value of this sum, a state with a lower value has relatively less energy.

Interestingly, atoms with the same outer shell structure also have a similar optical absorption or radiation spectra structure. This is because these spectra are caused by the transition of weakly bound external electrons to the levels the rules allow. The internal electrons do not participate in the changes since they are rather strongly connected to the nucleus of an atom.

Under the influence of the external environment, the structure of the optical spectrum of atoms changes. For example, this structure becomes more complicated when an atom is placed in a constant and uniform magnetic field. This is caused by the influence of the area on degenerate energy levels, which leads to their splitting and, as a consequence, to the splitting of spectral lines (*the Zeeman effect*). In this case, there are regular and abnormal splits; classical methods calculate the first, and the second does not. The influence of a weak magnetic field when the Zeeman splitting is minor compared to the splitting under spin–orbit interaction allows us to explain the simple and complex Zeeman effect.

Thus, considering the behaviour of atoms in the structure of human neurons enriches AI science by representing AI models while considering non-local, wave, and other effects.

7.1 Chapter Conclusion

- Neurons are made up of atoms, and their behaviour depends on the structure and behaviour of atoms; therefore, the understanding of thinking processes and their representation by AI system should consider the general subatomic structure of matter.
- Elementary particles and a photon are enough to build almost all the matter and fields, including the matter foundation of thinking processes; however, the photon, as it is believed, doesn't have a structure, which does not agree well with its ability to carry out spontaneous fluctuations.
- There is a non-classical idea that a single photon itself carries information about the source, for example, a star in space; to prove this idea, it is worth experimenting with recognising the difference between the spectrum of two photons which are radiated from two sources with the help of the double-slit effect.
- The similarity of the behaviour of an elementary particle and a word in human thoughts may be noticed; as a consequence, the thinking processes can be associated with a single photon flying towards the diaphragm and the screen behind it, which "does not know" where it will fall on the screen.
- Considering the behaviour of atoms in the structure of human neurons can accelerate AI development towards creating Strong AI.

References

1. Braibant, S., Giacomelli, G., Spurio, M.: Particles and fundamental interactions: an introduction to particle physics. Springer, Dordrecht (2012). https://doi.org/10.1007/978-94-007-2464-8
2. Bartels, R.A., Murray, G., Field, J., Squier, J.: Super-resolution imaging by computationally fusing quantum and classical optical information. Intell. Comput. **2022**, e0003 (2022). https://doi.org/10.34133/icomputing.0003
3. Moreau, P.-A., Toninelli, E., Gregory, T., Padgett, M.J.: Imaging with quantum states of light. Nat. Rev. Phys. **1**(6), 367–380 (2019)
4. Pawlowska, M., Tenne, R., Ghosh, B., Makowski, A., Lapkiewicz, R.: Embracing the uncertainty: the evolution of SOFI into a diverse family of fluctuation-based super-resolution microscopy methods. J. Phys. Photonics **4**(1), 012002 (2021)
5. Mangeat, T., Labouesse, S., Allain, M., Martin, E., Poincloux, R., Bouissou, A., Cantaloube, S., Courtaux, E., Vega, E., Li, T., et al.: Super-resolved live-cell imaging using random illumination microscopy. Cell Rep. Methods **1**(1), 100009 (2021)
6. Michler, P., Imamoğlu, A., Mason, M.D., Carson, P.J., Strouse, G.F., Buratto, S.K.: Quantum correlation among photons from a single quantum dot at room temperature. Nature **406**(6799), 968–970 (2000)

7. Raikov, A.: Cognitive semantics of artificial intelligence: a new perspective. In: Topics: Computational Intelligence XVII. Springer, Singapore (2021). https://doi.org/10.1007/978-981-33-6750-0

8. Puetz, S.J.: The infinitely fractal universe paradigm and consupponibility. Chaos Solitons Fractals **158**, 112065 (2022). https://doi.org/10.1016/j.chaos.2022.112065

9. Everett, H.: Relative state formulation of quantum mechanics. Rev. Mod. Phys. **29**(3), 454–462 (1957)

10. Harre, R.: Intern. Stud Philos Sci **4**, 2 (1990)

11. Dalela, A.: Quantum meaning: a semantic interpretation of quantum theory. Shabda Press, Kindle Edition, Bangalore, India (2012)

12. Raikov, A.: Strategic analysis of the long-term future needs of educational services. In: Proceedings 3rd World conference on smart trends in systems, security and sustainability. Roding Building, London Metropolitan University, UK. IEEE, pp. 29–36 (2019). https://doi.org/10.1109/WorldS4.2019.8903983

13. Einstein, A., Podolsky, B., Rosen, N.: Can quantum-mechanical description of physical reality be considered complete? Phys. Rev. **47**, 777–780 (1935)

14. Bell, J.S.: On the Einstein-Podolsky-Rosen paradox. Physics **1**(3), 195–200 (1964)

15. The BIG Bell Test Collaboration: Challenging local realism with human choices. Nature **557**(7704), 212–216 (2018). https://doi.org/10.1038/s41586-018-0085-3

16. Raikov, A.: Unpredictable threats from the malicious use of artificial strong intelligence. In: Pashentsev, E. (eds.) The Palgrave handbook of malicious use of AI and psychological security. Palgrave Macmillan, Cham, pp. 607–630 (2023). https://doi.org/10.1007/978-3-031-22552-9_23

Chapter 8
Dark Mind

There is a sizeable dark and still unknown part of the human neural brain. Hypothetically, the Universe also has a significant amount of still unknown dark energy and dark matter. Along with the quantum entanglement phenomena, this comparison may suggest a possible relationship between the brain and the Universe, including their dark parts.

The photon is still considered a phenomenon that has yet to be understood in the context of human mind behaviour. Even though photons carry light, there are still dark spots in understanding their behaviour. For example, photons of relic radiation under the influence of gravity create different spectra in different directions, which suggests the presence of dark energy or dark matter. The existence of dark energy or dark matter is being questioned, especially in the context of new research using the James Webb Space Telescope (JWST) [1]. However, we will not plunge into these doubts since our task is only to point out the possible physical relationship between the behaviour of the human brain and its distant environment.

Chapter 1 remarked, hypothetically, on a sizeable dark part of the human neural brain structure. This is the part that behaves unclear. The matching availability of dark brain volume and the dark energy and dark matter in the Universe seems mysterious; however, this matching can shed light on the still unknown sides of behaviour and the origin of various cognitive and cosmic phenomena, which can be used to develop the capabilities of AI systems (see Chap. 7).

Dark neurons are everywhere in the human brain. They behave passively; no situation at the entrance will force them to generate an impulse at the exit. They take away energy from the body to maintain their existence. At the same time, their purpose is not yet clear to classical science. Dark neurons incredibly rarely send impulses by themselves and not due to a cumulative reaction to many incoming impulses. This is almost all that is known about their functional behaviour and purpose. In this chapter, we pay attention to the dark components of the Universe in the hope of finding some analogies and maybe creating some idea of a relationship between the human brain and the Universe.

A. Raikov, *Photonic Artificial Intelligence*,
SpringerBriefs in Computational Intelligence,
https://doi.org/10.1007/978-981-97-1291-5_8

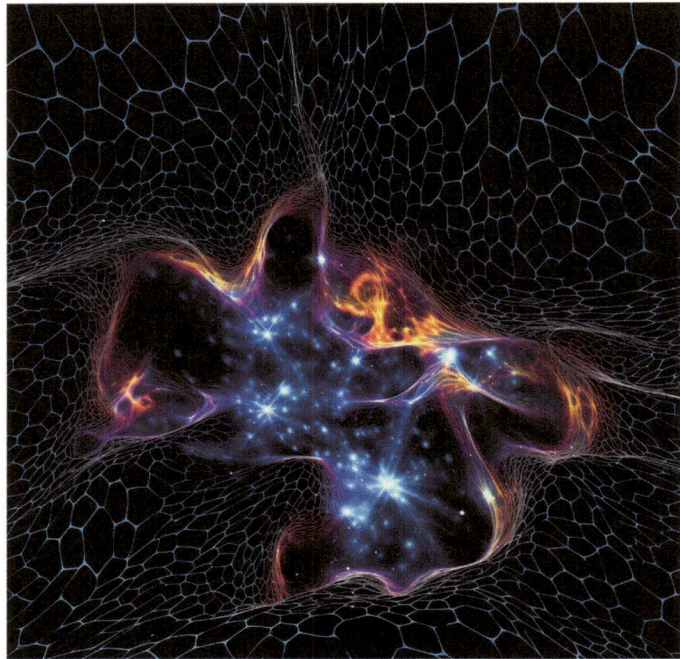

Fig. 8.1 Dark phenomena of the Universe (published with permission by Openverse Ltd)

Almost nothing is currently known about the structure of hypothetical dark matter and dark energy and the construction of the dark part of the human mind, deepening to the subatomic level and rising to the cosmic one. The dark phenomena of the Universe may be illustrated in various ways, including the use of generative Artificial Intelligence, the pictures of which can be found in the vastness of the Internet, or exclusive illustrations, such as the one shown in Fig. 8.1 (designed using the LLM by Openverse Ltd.

Over the past decades, about 50,000 scientific articles and books have been published on dark matter and dark energy research. In these works, the "dark" substance is considered primarily from a physical point of view, and their inter- action with the human mind is not affected. At the same time, it may determine the mystery of a person's subjective factor, mind and transcendental states of mind. However, in the Renaissance, the great sage D. Cardano, credited with creating the theory of probability and imaginary numbers, believed that planets and stars are not so far away from us and influence man's fate. The phenomenon of astrology is outside of science, although there are attempts to link science and astrology [2].

Let us turn to *dark matter*. Dark matter is a hypothetical component of the matter of the Universe, which, according to modern knowledge, so far only manifests itself by its gravitational effects and does not designate itself by other known types of physical interaction (electromagnetic, strong, weak). By its quantity, dark matter makes up about 26% of the matter of the Universe.

Its central core of the modern understanding of dark matter is the assumption of some unknown disturbing body or force that introduces minor distortions in the directions of the rays of light of stars. This phenomenon has helped to discover new planets, refine the organisation of stellar systems, predict and investigate black holes, etc. The absolute luminosity of large spiral-shaped cosmic structures was estimated by the mass and magnitude of the radiation, which helped to determine the distances to space objects and the scale of the Universe. The results of observations and measurements were influenced by the structure of the object (nebula, star, etc.), luminosity, and lifetime.

The intergalactic glow was studied depending on its concentration around galaxies of different sizes; it was shown that it is most evident around giant galaxies. At the same time, the contribution of this glow to the total mass of matter in the Universe is relatively insignificant; that is, this component is not enough to explain the total mass of the Universe. It was also noticed that the images of nebulae intersect, on which the effect of gravitational lensing was justified. This helped provide the most straightforward and accurate determination of the masses of galaxies. For example, one can find illustrations of the Bullet Cluster in [3], which show the accumulation of high-temperature gas in intergalactic space after the collision of galaxies, as well as the distribution of dark matter that covers both clusters.

Photons from distant stars, galaxies, and their clusters propagate in the space of gravitational fields created by groups of celestial bodies. Therefore, in classical terminology, the particle flies along a hyperbolic trajectory, provided that its velocity is higher than the velocity of the massive source of this gravitational field that gave rise to the particle. However, due to the curvature of space, the light rays flying on different sides of this source, after deflection, can converge again at one point. A massive body with its immediate surroundings can be viewed as a lens that changes the path of light beams.

A quantum analogy also suggests itself in the context of studying the brain's quantum particle behaviour. In one of the quantum interpretations, a quantum particle can have an infinite set—a superposition—of paths, of which only one of the paths is fixed at the moment of observation (measurement). In the space case, by a telescope, the observer sees the paths of light rays from the source (stars, galaxies, etc.), which arose due to the collapse of multiple states.

By measuring the parameters of these rays and applying inverse problem-solving methods, it is possible to restore the mass of the gravity source, even if this mass may be invisible. A lot depends on solving this problem from the size and mass of the gravity source that the rays band around.

Unlike dark neurons, the composition of dark matter is unknown; only the influence of dark matter on the organisation of galaxies and the change in the trajectory of light rays from distant stars is shown. There are theoretical hypotheses regarding the composition and approximate mass of dark matter. For example, a complete analysis of the influence of dark matter, hypothetically consisting of massive neutrinos, on the evolution of the Universe and processes in the early Universe was given by the

Zeldovich group [4–6]. This helped, in particular, to explain one of the paradoxes of cosmology associated with the discrepancy between the available cosmological model and the observed data.

The transitions of the hyperfine structure of the neutral hydrogen level are associated with the electron and proton spin. The hyperfine structure is determined by tiny shifts in degenerate energy levels and the resulting splitting of atoms, molecules, and ions at these levels due to the electromagnetic interaction between the nucleus and electrons. With such a transition, a radio line with a wavelength of 21 cm (1420 MHz) appears in the spectrum of signals from Space. This is called the neutral hydrogen line. It helps to assess many uncertainties associated with the unknown composition of interstellar space that this line emits and absorbs, as well as the probability that all hydrogen is present in the form of neutral molecules. It is noted that large spiral galaxies are surrounded by a halo (secondary glow) of neutral hydrogen. This feature of interstellar space demonstrates that "inconspicuous trifles" can cover and help to understand massive objects and phenomena.

There are also some "inconspicuous trifles" in the neural structure of the human brain and body that are ignored when creating modern AI systems in the form of neural networks. They appear at once when scientists begin to consider the subatomic structure of neurons. For example, there are the "darkness", non-local quantum effects, and fluctuations in neuron behaviour. Let us also note that the rhythms of most human body systems lie in the infrasound range at different times of the day: delta brain rhythm (sleep state)—0.5–3.5 Hz, alpha brain rhythm (rest state)—8–13 Hz, beta brain rhythm (mental work)—14–35 Hz.

It is shown that the initial perturbations of the density of matter in the early Universe generate characteristic inhomogeneities of the relic radiation that appeared 330 thousand years after the birth of the Universe (according to the hypothetical Big Bang model). The paradox is that these initial elements of matter, baryonic matter, would not have had time to shrink under the action of their gravitational forces so that by now, galaxies, galactic clusters, nebulae, stars, and planets would have formed.

Invisible dark matter can explain and resolve this paradox since it began to contract earlier than ordinary (baryonic) matter. Consequently, the growth of perturbations in matter density started much earlier. On the scale of galaxy clusters, first, there was an increase in perturbations in the density of dark matter. Then, these regions, under the influence of gravitational forces, began to attract baryonic matter to themselves for the subsequent construction of cosmic formations of various sizes. At the same time, the density of dark matter is much less than that of baryonic matter.

Today, we can assume that certain parts of the Universe's matter are in a form unknown to classical physics, resembling a substance of heavy, cold particles. When we talk about cold dark matter, we mean models whose components are weakly interacting particles with such mass and annihilation cross section, which had allowed them to get out of equilibrium in the early epochs of the Universe birth, when their density and relict prevalence had become corresponding to hypothetical dark matter. Other particles, such as neutral bosons, were not initially associated with dark matter.

The future process of dark matter cognition may help better understand the poorly studied phenomenon of the human mind. The model with dark matter currently explains cosmological and astrophysical phenomena, including constructing space objects from initial fluctuations, considering the influence of disturbances and the primary heterogeneity of matter.

There are a number of theoretical proposals for estimating the mass and density of dark matter—for example, prof. M.V. Sazhin (see his work about cosmic strings [3]) proposed to develop an AI method to estimate dark matter characteristics precisely. In a classical statement—without AI—he suggested using the virial theorem to the movements of stars. So, the kinetic energy E is a quadratic function of velocities. According to the Euler's theorem on homogeneous functions, one can write the equation:

$$\sum_i \frac{\partial E}{\partial v_i} v_i = 2E. \tag{8.1}$$

After entering the impulses p_i into this equation, which, according to the canonical equations of classical mechanics, can be expressed as

$$\frac{\partial E}{\partial v_i} = p_i; \tag{8.2}$$

we will get the following:

$$2E = \sum_i p_i v_i = \frac{d}{dt}\left(\sum_i p_i r_i\right) - \sum_i \frac{dp_i}{dt} r_i. \tag{8.3}$$

Let's average this equation over time. If the function $f(t)$ being averaged in this case is a time derivative of a bounded function $F(t)$ (i.e. a function that does not accept infinite values), then its average value vanishes:

$$\overline{f} = \lim \frac{1}{\tau}\int_0^\tau \frac{dF}{dt} dt = \lim_{\tau \to \infty} \frac{F(\tau) - F(0)}{\tau} \to 0. \tag{8.4}$$

Suppose that a system of particles moves in a finite region of space and with velocities that do not turn to infinity; then the value:

$$\sum_i p_i r_i \tag{8.5}$$

will be bounded, and the average value of the first term in Eq. (8.3) vanishes. In the second term, the derivative of momentum:

$$\frac{dp_i}{dt}, \tag{8.6}$$

According to Newton's equations, we replace it with the derivative of the gravitational potential:

$$-\frac{\partial \phi}{\partial r_i}.$$

(8.7)

We will get the following:

$$2E = \sum_i r_i \frac{\partial \phi}{\partial r_i}.$$

(8.8)

The gravitational potential is a homogeneous function of the modulus of radius vectors r_i; according to Euler's theorem, this equality passes into the desired relation:

$$2\overline{E} = -\overline{\phi}.$$

(8.9)

Knowing the velocity of stars from satellite measurements, it is possible to calculate the average kinetic energy in a specific volume and, according to (8.9), calculate the average potential in this volume.

Further, after applying the d'Alembert operator to the gravitational potential, the value of the total density of matter in this volume will be obtained:

$$\Delta \phi = 4\pi G \rho.$$

(8.10)

This makes it possible to estimate the density of matter, represented by baryons (stars), and, after subtracting one from the other, determine the density of dark matter.

This proposal is based on several assumptions that must be worked out during research. Thus, it follows from Euler's theorem for homogeneous functions that the eigenfunctions of the operator introduced by this theorem are homogeneous functions. However, there is no certainty that the space containing dark matter is homogeneous. The space can also be an inhomogeneous curved and a curvature tensor. There may even be phenomenological variables characterising the system under consideration, fluctuations, chaos (information) derived from the chaos of the internal and the chaos of the exchange of the system with the external environment, which is not invested in quantitative metrics. The assumption that a system of material particles moves in a finite region of space also needs to be clarified because the finiteness of space is not apparent, and the effects of quantum non-locality are manifested.

Along with the above aspects, it is also necessary to use and develop an appropriate mathematical apparatus for modelling the solution of the human brain and cognition, as well as cosmological and hydrodynamic problems. The physics-informed neural networks (PINN) method can help to do this [7]. PINN approach can dramatically accelerate the solution of direct and inverse problems related to processing various flows, for example, blood, cerebrospinal fluid, etc., in the hollow tubular structures (vessels, subarachnoid spaces), which create the needed means for the work of the human brain's neural structure, considering the properties of the fluid and the construction of the vascular wall, and the longitudinal vorticity in the circulatory system.

Now consider *dark energy*. The organisation of dark energy is most often regarded as a negative gravity that patches up gaps in classical theory. For the theory to work, it must be assumed that this energy occupies about 70% of the total energy in the observable Universe.

Negative gravity pushes space objects apart, which baryon gravity cannot do. Thanks to dark energy, the Universe expands, and expands with acceleration, which, in turn, is sometimes considered not as natural acceleration but as a slow decrease in the speed of cosmic expansion due to braking by the forces of ordinary gravity. It is also believed that this energy distributes evenly in space, and its density does not depend on time.

Of course, there are other opinions. For example, new research using the JWST calls into question the idea of the Big Bang and the expansion of the Universe [1]. The cosmological studies with JWST have shown the presence of distant galaxies that are not moving away as quickly as many others do.

There are various other explanations for the presence of dark energy, to the point that it simply does not exist. For example, there is no need to introduce this concept if you look at our Universe as if from the outside and consider the world infinite, fractal [8], and the Universe (with a diameter of 92 billion light years) is its small piece (after all, the Universe is perhaps expanding somewhere).

Often, the phenomenon of dark energy is explained by the idea of a cosmic vacuum, which creates an anti-gravity field. At the same time, such an idea encounters difficulties of explanation; the main one is that the density of a vacuum changes with the cosmic expansion because as any object expands, its density must fall. To compensate for this limitation, the idea of vacuum is modified, making this density variable. That is, the density of the vacuum changes, which looks very paradoxical because the essence of the vacuum is that there is nothing in it (as known, this is not the case). However, an explanation can be found for this if we assume that the expanding space does work on dark energy, which leads to an increase in this energy [9].

The form of representation of dark energy in theoretical models is the cosmological constant, which can be viewed by reflecting the influence of both positive and negative masses of empty space and scalar fields that dynamically reflect changes in energy density in time and space. This constant may vary from zero to very large.

Sometimes, a particular quintessence is put on the role of dark energy, which corresponds to a value of the cosmological term (< 0) and depends on time. To explain dark energy, the Sachs-Wolf effect (simply an expression of energy conservation) is used. This effect reflects the impact of a changing gravitational field on photons of relic radiation, leading to the anisotropy of the spectrum of this radiation.

The phenomenon of dark energy is very suitable for the name of the phantom field, which is well associated with fantastic and exotic problems. However, if the dark energy is a phantom with an infinitely increasing density, an unenviable fate awaits our Universe. The Universe's expansion will constantly accelerate, and the Big Bang model will be replaced by the model of a Big Gap [9].

The connection of dark matter, dark energy, and dark brain is hypothetical, and it borders, to some extent, with science, fantasy, and romance. Objections are formulated, only some arguments are given in defence of this hypothesis, and there is a dispute, the end of which is not visible in the near and long term.

At the same time, physicists mainly participate in this discussion. The idea of comparing the phenomena considered with the behaviour of a somewhat mysterious phenomenon of nature—mind, shadow neurons in the dark brain—is practically not considered in this discussion. However, it seems fruitful to consider such an analogy or connection, which can lead to new information about the dark structure of reality, including the Universe, mind, and thinking.

The connection between micro and macroworlds has already been proven, e.g. in the form of the quantum entanglement effect. The relationship between the structures of the brain and cosmic matter is too small on the probability scale. If the Big Ban model and the entanglement quantum effect are actual, then all Universe's particles are connected in pairs. Still, the probability of connection of any individual's brain's atoms cannot be more than 10^{-56}, considering the approximate estimates of particles in the brain and the Universe.

However, this connection can be more fundamental. For example, the continuous paradigm of mind and thinking processes makes the hypothesis of the Big Bang model, the finiteness of the Universe, and the local semantics of AI models very controversial. These hypotheses reduce the continuity of the thoughts and mind to the digital signals in AI systems development. In contrast to the digital approach, with the help of photonic techniques, continuous signals can be processed without their digitalisation. This will allow the non-local cognitive semantics of AI models [10] to be taking into account more deeply, and computational processes will be significantly faster.

The comparative research of processes in cosmology and astrophysics with processes in mind and thinking can construct some analogies between the behaviour of the human thinking processes and processes in advanced AI in the form of photonic ones.

8.1   Chapter Conclusion

- Processes in the Universe and the human mind are influenced by hidden (dark, shadow) factors and phenomena, the nature of which is still poorly explained and does not fit into the classical physical and mathematical paradigms, is phenomenological (cosmological constant, ray deflections, dark matter and dark energy, dark neurons).

- Hidden (dark) phenomena in the human brain and the Universe, making up most of them, have a specific localisation and a fuzzy boundary separating them from the known world and phenomena related to the behaviour of objects of baryonic matter.

- Analogies of the organisation of matter at the quantum and cosmic levels may be suggested in terms of the hidden (dark) parameters, such as dark neurons, dark matter and dark energy, shadow photons, and superposition of quantum states.
- A comparative study of the "dark sides" of the Universe and the human mind, as a whole, can help us understand the structure of reality and the history of the Universe's birth.
- Photonic techniques can be used to prevent reducing the continuity of thoughts and mind to the digital signals in AI systems.

References

1. Khlopov, M.: Recent advances of Beyond the Standard model cosmology (2024). https://doi.org/10.48550/arXiv.2401.04735
2. Brooks, M., Cardano, G.: The Quantum Astrologer's Handbook: a history of the Renaissance mathematics that birthed imaginary numbers, probability, and the new physics of the universe. Scribe Publications, Melbourne (2017)
3. Sazhin, M.V., Khovanskaya, O.S., Capaccioli, M., Longo, G., Paolillo, M., Covone, G., et al.: Gravitational lensing by cosmic strings: what we learn from the CSL-1 case. Mon. Not. R. Astron. Soc. **376**(4), 1731–1739 (2007)
4. Zeldovich, Yu.B., Syunyaev, R.A.: Astronomical consequences of the neutrino rest mass. I. The universe. Lett. Astronom. J. **6**(8), 451–456 (1980) (in Russian)
5. Doroshkevich, A.G., Zeldovich, Yu.B., Syunyaev, R.A., Khlopov, M.Y.: Astrophysical implications of the neutrino rest mass. II. The density-perturbation spectrum and small-scale fluctuations in the microwave background. Lett. Astronom. J. **6**(8), 457–464 (1980) (in Russian)
6. Doroshkevich, A.G., Zeldovich, Yu.B., Syunyaev, R.A., Khlopov, M.Y.: Astronomical consequences of the neutrino rest mass. III. Nonlinear stage of perturbation development and hidden mass. Lett. Astronom. J. **6**(8), 465–469 (1980) (in Russian)
7. Raissi, M., Perdikaris, P., Karniadakis, G.E.: Physics-informed neural networks: a deep learning framework for solving forward and inverse problems involving nonlinear partial differential equations. J. Comput. Phys. **378**, 686–707 (2019). https://doi.org/10.1016/j.jcp.2018.10.045
8. Haitun, S.D.: The hypothesis of the fractality of the universe: origins. Grounds. 24 investigations. M. Lenand, p. 336 (2018) (in Russian)
9. Rubakov, V.A.: Dark energy in the universe. Trinity Var. **14**(258) (2018) (in Russian)
10. Raikov, A.: Cognitive semantics of artificial intelligence: a new perspective. In: Topics: Computational Intelligence XVII. Springer, Singapore (2021). https://doi.org/10.1007/978-981-33-6750-0

Chapter 9
Material Synthesis

A unique material must be created for quickly rewritable 3D holographic memory for the photonic AI. A finite number of images must be rapidly written and re-written by laser beams in one location (point, dot, spot) of the memory body. There is no such material, but some approaches give a chance to synthesise one.

9.1 3D Holographic Idea

The image can be written in holographic memory by laser beams as themselves or as a Fourier image. The latter requires less physical space for storage points and less susceptibility to imperfections in the material. The requirements for such material are listed in Chap. 5.

In the required material for a rewritable 3D holographic storage device, a coherent modulated beam of light (photons) makes a record. The photon is stable, but the medium for recording is unstable relative to the photon. Real photons form electromagnetic waves: radio waves, ultraviolet and infrared radiation, ordinary light, X-ray radiation, and gamma radiation.

Spin-photon interfaces may be essential connections between stationary quantum particle states of photonic material that can, for a long time, store the information and flying photons. The basic idea of spin-photon interfaces is to entangle the spin degree of freedom of the quantum particle states with the energy or polarisation of the flying photon.

In electromagnetic interaction, the difference in the signs of particle charges in the photonic material of a small volume (point, dot) leads to mutual compensation and neutralisation of the total charge. Since the desired material consists of elementary particles connected by electromagnetic interaction, the fundamental level of particles and electrodynamics should primarily form the phenomenological and physical basis

A. Raikov, *Photonic Artificial Intelligence*,
SpringerBriefs in Computational Intelligence,
https://doi.org/10.1007/978-981-97-1291-5_9

for synthesising the necessary (quantum-stable) material for a rewritable holographic storage device.

At the same time, it is worth noting that even a vacuum, associated with the absence of matter, is not empty and is an unstable structure. A vacuum is a state of a quantum field with minimal energy, in which there are no excitations, particles or quasiparticles, or a medium whose tension is equal to the volumetric energy density. A vacuum always has energy fluctuations, following the Heisenberg uncertainty relation. The vacuum is characterised by polarisation, which includes generating a particle-antiparticle pair or spontaneous generation of an electron–positron pair. Thus, searching for and creating material with "absolute" stability makes no sense. The ideal required material will provide maximum stability at an energy potential exceeding the potential of a vacuum medium of a specific volume. It will be commensurate with the possibility that makes up the total potential of the fundamental particles of this volume.

The wave function describes the state of a quantum system. The Schrodinger equation allow one to predict its behaviour before and after a certain moment. The solution of the Schrodinger equation is stable according to the initial data; it is also allowed that some errors introduced into the initial data remain unchanged over time.

However, this stability instantly disappears if the system is subjected to measurement or external influence, such as photonic. The wave function exhibits a probabilistic nature, which is characterised by irreversibility. Irreversibility accompanies the uncontrolled interaction of particles, which is a consequence of the fact that the environment external to the quantum system contains many particles. Sometimes, the result of measuring the wave function can be predicted unambiguously. For the same state of a system with the same wave function, another parameter can be selected, the measurement of which will no longer be unambiguously predictable, for example, parameters related to the uncertainty relation.

For a general understanding of the process of synthesis of required materials, it suggests making some simplified analytical solutions based on the methods of classical physics. For example, the behaviour of a classical system (material) can be described as the evolution of a set of probabilities in the phase space of particle coordinates and pulses. For unstable systems over long periods, the reliability of such descriptions is doubtful. In quantum theory, in addition to phase probability, the probability of amplitudes appears as complex numbers whose modulus square sets the probability of a state. In this case, the probability is mutually uniquely determined by the probability amplitude module, which stores data on the phase of the probability amplitude.

Consider a molecule, the basis for constructing the desired holographic material. A molecule combines several atoms connected in a particular geometric order. A molecule is a sample of a relatively stable physical structure; various atoms in it are united strictly according to specific rules, which depend on the properties of atoms and the distribution of electrons in orbits.

A molecule includes links, and its properties strongly depend on changes in the number of repeating links in the molecule and the nature of the end groups. The bond between atoms in a molecule is created by the "socialisation" of electrons.

Each connection in the molecule is a divided pair of electrons. The number of links an atom can form with other atoms depends on the number of electrons that this atom shares with other atoms. These properties of particles obey rules such as the carbon atom having four links, nitrogen having three, chlorine having one, and hydrogen usually having one. Some atoms can form several bonds with others, but no more than three.

The unique role of carbon in the synthesis of materials is due to the unique ability of its atoms to form bonds with each other—after all, it has more than all other atoms to bind. Carbon is the most "friendly" of all elements; it has a moderate ability to accept electrons from other atoms and quickly gives up its electrons for connection. Therefore, it is carbon and its derivatives in the form, for example, graphene, that may be of interest for synthesising the required material with modelling approaches.

When modelling material synthesis processes, the initial coordinates and particle velocities can be set based on X-ray spectroscopy and nuclear magnetic resonance data. The values of the parameters of interatomic interactions are determined from the condition of correspondence of the spectral, thermodynamic, and structural characteristics of molecules calculated by potential and measured.

Physical-chemistry reactions and processes, such as chaotic diffusion of matter transfer, related to the pressure of external forces (environment, gravitational and electromagnetic fields) occur in molecular behaviour. The classical approach to modelling molecular dynamics processes involves solving the equations system of many molecules, which is well known. The critical computational problem for such a model is that molecular dynamics can be nonlinear depending on force fields. Then, most of the calculations will be spent on force field calculations.

The molecular dynamics equations systems may be too complex for calculations; there are many ways to decompose and solve them, especially by supercomputers with parallel programming. However, it should be noted that this requires a relatively large amount of computer time and energy. The two most popular methods are molecular decomposition and spatial decomposition. The third method, rare, considers the decomposition of forces.

Let us consider modern approaches to creating required materials: AI technology based on genetic algorithms and the classical physical-chemistry approach, represented by graphene and quantum dots.

9.2 AI Modelling

It has become possible to create new materials by exploiting modern AI with molecular dynamic modelling methods for crystal structure prediction, identifying metastable phases with low-lying local minima assessing [1]. The inputs for the modelling could be the sets of different types of subatomic particles and molecules with their characteristics. A variety of materials (photonic crystals, crystalline surfaces, quasicrystals, graphene, etc.) can be handled by crystal structure prediction with their features.

Different structures can be generated randomly by relaxing each to their lowest-energy local minimum. It leads to a set of local minimum-energy configurations. Databases such as the Inorganic Crystal Structure Database are resources that collect observed structures of inorganic materials. This source contains more than 210,000 data entries and covers the literature from 1913; around 12,000 new forms are added yearly, and its content is modified due to continuous quality assurance. Even the older data is dynamic. All crystal structure data are available, including unit cell, atomic parameters, molecular formula, weight, Adventrx Pharmaceuticals formula, mineral group, etc. However, the number of structures in this database is small compared to those generated based on the first-principles methods.

Data mining approaches for crystal structure prediction based on exploratory algorithms help generate a counterintuitive new structures beyond existing databases. These approaches meet challenges, including working with large structures considering disorder and predicting possible metastable structures and their properties.

Density functional theory (DFT) and modern supercomputing technologies can help to assess and relax many structures by the Born–Oppenheimer meaning, which is the mathematical assumption and approximation in molecular dynamics that the wave functions of atomic nuclei and electrons in a molecule can be treated separately. Crystals may contain hundreds of atoms in the unit cell, which is the structure and characteristics of which modern supercomputers cannot calculate and predict using the first-principles approach. To reduce the structural optimisation and energy evaluation procedures, interatomic force fields are employed on a high level of abstraction; more accurate evaluations are used for the short list of candidates by the DFT.

Papers [2, 3] demonstrate a non-empirical optimisation method based on restructured classical chemical space, energy filtering, and evolutionary algorithms to predict synthesisable materials with optimal properties. This method aims to find materials with the required properties of the highest hardness, the lowest dielectric permittivity, etc. However, some problems remain unsolved: the accurate prediction of a material with optimal properties among an infinite number of crystal structures cannot be constructed, exhaustive screening of such a large set by computer is unrealistic, data mining statistics show that the existing databases are incomplete, data mining as an autoregression method cannot find fundamentally new crystal structures.

The crystal structure and properties are determined by data on stoichiometry, atomic size, polarizability, and electronegativity of atoms [4]. Discovering materials with optimal properties requires an adequate organisation of the chemical space, e.g. if elements' atomic numbers order the area, it is not fit for global optimisation. The paper [1] suggested a redefined chemical scale or the Mendeleev number using a non-empirical way that clarifies the physical meaning of elements by using such chemical properties of the atom as its size and the Pauling electronegativity—the tendency of the atom to attract electrons.

It should be noted that these approaches have been used to create stable superhard materials, superconductors, and organic materials. These tasks do not cover the problems of creating photonic crystals for the rewritable holographic memory with many stable states. However, computer crystal structure prediction methods can be used to solve the issue of synthesising photonic crystals with the required parameters.

Paper [5] shows that spin-coated photopolymers on glass plates ensure surface uniformity. Volume holographic transmission gratings with a diffraction efficiency of 80% could be stored in photopolymers of 100 μm thickness. Such efficiency is achievable for gratings recorded with energy 10 mJ/cm^2.

Paper [6] studies the diffraction effect when atoms replace electrons under quasi-resonant laser radiation. In this direction, such regimes as the Raman-Nath and the Bragg were used [7]. The former covers an atom-light interaction when the scattering process creates different atom momentum states. A long interaction time and a single diffraction order were allowed for the latter. The experimental study of the connection between beam splitters, diffraction efficiency, and phase showed the required size of the momentum basis for calculations of the diffraction effects. It confirmed the relationship between diffraction phases and the non-adiabatic losses.

Paper [8] describes a holographic memory disc on the base of a hybrid opto-electronic correlator (HOC) [9]. The HOC architecture could detect matches during the shift, scale, and rotation invariant correlation [10]. The holographic memory disc included more than a thousand stored images; inputs of the HOC are shifted to produce correlation signals. The polar Mellin transform pre-processor was also used, which helps convert images using the log-polar coordinate function. The objects detected in shift, scale, and rotation invariant manner were then numerically simulated by digital computing to validate the concepts.

A classical two-beam approach with a spatial light modulator is used to project and write the image into the storage in this system. A holographic memory disc replaced the spatial light modulator to get a zero-latency property of the processes. At the same time, the proposed approach uses a field-programmable gate array[1] to implement its technology, that is, a logic-integrated circuit programmable on a digital computer connected with an optical part by USB (Universal Serial Bus). This can be useful for creating external non-erasable data storage or memory but imposes significant time limits when used online without the possibility of overwriting data.

The mentioned system ensures a write-once-read-many holographic storage media [11, 12]. The phenanthrenequinone doped with a photosensitive dye poly (methyl methacrylate) was used to create the material that supports different applications, such as data storage and wavelength filtering. The thickness of such holograms is a few millimetres; it works in the Bragg-type regime—when the pulse duration is sufficiently long to suppress diffraction into spurious orders. The laser beam exposed to the material forms oligomers[2] by increasing the material particles' energy state. It can generate a non-uniformity of the material's refractive index[3]

---

[1] A type of circuit that can be programmed or reprogrammed after manufacturing.

[2] An oligomer is a molecule that consists of a few repeating units, including smaller molecules and monomers. A monomer is a molecule that can react with other monomer molecules to form a larger polymer.

[3] The refractive index is a dimensionless number that indicates the light-bending ability of the optical medium.

with its spatial modulation. Due to this property, the images may be imprinted onto the holographic material. The value of the refractive index yields 100% diffraction efficiency. Overlapping holograms at a single location can be written due to Bragg conditions, which, due to the diffraction effect, make different angles or wavelengths for coherent scattering beams [14].

Paper [15] compared the dyes/pigments with structural colouration. The latter is characterised by finer tunability, sustainability, and durability. Some high colour purities and designs were demonstrated based on the photonic crystals, plasmonic nanostructure, and thin-film interferometer. This mentioned paper is devoted to the issue of escaping photonic material from the static state fixed by initial structures. For example, a tunable photonic characteristic has phase change material due to its significant variations in optical properties in both imaginary and real parts of the complex refractive index. Its compatibility with complementary metal-oxide semiconductors, e.g. Ge–Sb–Te, ensures different switchable photonic applications, such as neuromorphic photonic memory, terahertz photonic devices, etc. Their optical parameters stem from the changes in their atomic bonding configuration in the metal–insulator transition media. Introducing free carriers in metallic parts causes significant losses due to their absorption.

The material's refractive index in its real part may be changed from high to low, while the imaginary part (extinction coefficient)—is from low to high. An optical loss limits the range of the optical wavelength regime. To unlock such limits, new material compositions have been discovered by introducing dopants into the active materials [16].

Aside from the pulse-code modulation approach, which is used to digitally represent sampled analogue signals, diverse optically active materials have been used to manipulate the photonic responses with appropriate refractive index and tuning range. Such materials can be as follows:

- Thermal energy-driven phase-change-transition materials.
- Chemical/electrochemical-reaction-based refractive index tuning in conducting polymers, dielectric, or metals.
- Electrical energy-driven dynamic birefringence material.
- Electrical/optical stimulation for free-carrier pumping in dielectric or semiconductor, resulting in modulation of the complex refractive index.

In the nanoscale of matter construction—from 0 to 100 nm—particles can interact, self-organise, and self-develop. Particle aggregates can maintain stability well. As noted above, the most stable particle is carbon because its atom has the most valence bonds compared to other atoms. Graphene occupies a special place in carbon structures.

9.3 Graphene Approach

The interatomic bond can be weak or strong. The 3D carbon allotropes demonstrate a weak van der Waals interlayer π-bond connecting the layers of matter at a distance of 0.335 nm. This is a bond when atoms come close to almost touching their external electron clouds, which induces charge fluctuations and leads to an undirected attraction. These interactions decrease in proportion to the sixth power of the distance.

An example of a strong bond is demonstrated by a two-dimensional (single-layer) allotropic modification of carbon, which forms the so-called *graphene*. The graphene layer is a hexagonal lattice with edges of 0.142 nm, in the nodes of which there are carbon atoms. In this case, each atom is connected to three neighbouring atoms by covalent chemical bonds, and the fourth valence electron is included in the conjugated system of graphene. Initially, it was believed that such 2D crystals were thermodynamically unstable. However, it was later shown that graphene stability is possible due to thermal fluctuations in the form of creating a graphite surface with a height of 0.5 nm [17].

A stable graphene film on an oxidised silicon substrate appeared much later [18, 19]. It turned out that a graphene film can be obtained using conventional adhesive tape. Graphene has many valuable properties; for example, light and durable materials for any vehicle can be made from plastic with the addition of graphene. Other 2D materials have already been applied; silicene is a two-dimensional allotropic silicon compound compatible with semiconductor materials [20]. Current carriers in silicene behave like particles without mass and move at a speed ten times less than the Fermi velocity, 300 times less than the speed of light characteristic of current carriers in graphene.

The properties of nanostructures stacked on a graphene substrate have been studied through DFT calculations [21]. It shows the critical role of edges in binding strength and interfacial electron coupling. A space-sampling method was used to evaluate the possibility of realising planar carbon allotropes [22] by assessing the frequency of occurrence of an allotrope. Two allotropes are proposed by plotting the occurrence frequency against the total energy.

The study of graphene properties is carried out using methods of quantum electrodynamics. The quasiparticles resulting from the collective interaction of electrons in a 2D crystal, called Dirac fermions, behave like electrons without mass, like relativistic particles with zero rest mass, although they move at Fermi velocity. The Dirac equation, as opposed to the Schrodinger equation, describes the behaviour of such particles.

Graphene made it possible to carry out relativistic modelling of the Klein paradox [23], which arises when solving the problem of tunnelling a quantum particle through a relatively high potential barrier. This barrier is entirely transparent for relativistic electrons. The solution of the Dirac equation made it possible to determine the coefficient of the passage of a particle as a function of its angle of incidence. It is shown that the probability of a particle passing through such a potential barrier in graphene at certain angles of inclination may cause resonances and reflections to occur [24]. This paper shows that the effect can be tested in a simple condensed-matter experiment using graphene electrostatic barriers.

The explanation of quantum field and particles' behaviour can be provided by quantum field theory. The Dirac equation describes the evolution of a quantum field over time, in which both particles and antiparticles are present. Pairs are born in strong fields, including particles formed behind the barrier. Computer simulation has shown that the electron is ultimately reflected from the barrier, and electron–positron pairs are created in the barrier [25]. This effect was identified with the Dirac equation, and the trembling motion was seen in a modern high-resolution microscope, having come to understand the essence of the minimum conductivity in graphene observed even at zero concentration of current carriers.

Phenomena that go beyond the classical concepts of physics observed in graphene are also interpreted by using methods of quantum electrodynamics and condensed matter physics. For example, such a combination of approaches helped to improve understanding of the mechanisms of the quantum Hall effect, the production of particles by Schwinger, particle interactions under the influence of Casimir forces, etc. [26].

Note that the quantum Hall effect is a quantisation of the transverse conductivity of a 2D electron gas at low temperatures in strong magnetic fields; unlike the classical Hall effect, in which the transverse conductivity changes monotonically, the quantum Hall conductivity changes stepwise with increasing magnetic field strength and electron concentration. This effect can help construct devices to change the spatial position of the light beam—optical deflectors.

A semiconductor material with a graphene channel can be a basis for constructing an infrared photodetector due to graphene's appropriate particle transport properties. Plasma oscillations (Dirac plasmons) also occur in graphene, which strongly interacts with THz-oscillations, forming quasiparticles that provide such interaction. In graphene, free charge carriers move only in two dimensions, and Dirac plasmons can exist at low frequencies. If the millimetre radiation has a wavelength from 0.03 to 1 mm, and the width of the graphene tape can be no more than 1 micron, we can obtain resonances in the THz range. Thus, graphene is suitable for creating photonic materials THz chips and controlling plasmons [27]. It allows us to get negative refraction of light, change the direction of the rays, and observe the reverse Doppler effect when interacting with fields.

Obtaining the quantum state of matter at temperatures of the order of nano-Kelvin (Bose–Einstein Condensate) may help build material for a photonic AI. The thin medium can reduce the speed of light down to zero. The ability to transmit coherent light of a certain length (narrow peak) through an opaque medium can be achieved using electromagnetically induced transparency. To do this, the powerful pumping field of the medium must be tuned to the difference in atomic energy levels of quantum transitions, which creates some destructive interference between the absorption paths.

Separately, the possibility of graphene becoming a superconducting material is worth noting. Graphene, consisting of two atomic layers rotated relative to each other by an angle of $1.1°$ [28], possesses this property. This property can accelerate and reduce losses in signal transmission processes in PAI components that provide contact between optics and electronics.

With graphene, compact electro-optical devices are already being developed—for example, high-finesse ring resonators or high-speed phase modulators for large-scale applications, including phased arrays, neural networks, and optical communication systems. Modern photonic materials for phase modulators exhibit low refractive index change. The index change induces a shift in the resonance wavelength of the response. However, these modulators require high index change. However, phase modulators are limited by their scaling capabilities due to the impossibility of getting a good trade-off between system length and optical loss; high insertion loss is associated with the phase change. This loss arises due to the significant mismatch between the difference in the real and imaginary parts of the index.

Conversely, the high-finesse resonators can have low insertion loss and minimal transmission variation due to modulation of the real and imaginary parts of the refractive index. For this, a hybrid platform has been combined on the base of a low-loss ring resonator using graphene technology and electro-refractive material [29]. To enable simultaneous tuning of the index's real and imaginary parts, the authors of this paper utilise a stack of tungsten disulphide and graphene integrated into a low-loss dielectric ring resonator. The hybrid platform enables the achievement of phase modulation at several GHz.

9.4 Quantum Dots

The writing of many images by the modulated laser beam in a single local point of photonic materials (is also known as angle multiplexing or angular multiplexing) can be ensured by the ability of a group of atoms, or quantum particles, to save stable different structure states corresponding to these images. The possibility of steady formation, transport, and stabilisation of multiple entangled quantum states has been shown in some physical systems [30–32], such as cavity quantum electrodynamics, atoms and molecules, and quantum dots (QD). It is an artificial entangled state of quantum particles with semiconductor properties. Their size may be less than the wavelength of light (380–780 nm) and is from 2 to 10 nm.

The creation of entangled states combined with relaxation rates control of quantum systems by weak resonance effects is already used in practice, including quantum-optical neural networks. The conditions for ensuring multi-photon resonances, the induction of stationary entanglement in two quantum bits, and its control by external fields were already found and shown. It is accounted that single-photon states that are immune to decoherence, are under control, stored, and detected are the basis of creating fault-tolerant photonic quantum computers; single photons and entangled photon pairs are the main components in photonic quantum technology [34, 35]. However, photons cannot be emitted deterministically in a classical way, and the number of pairs per pulse is characterised by statistical behaviour.

A non-classical approach with light sources like QD can emit a single photon and entangled photon pair deterministically. Such quantum is emitted under experimental conditions in the radiative recombination process [36]. Compared to other quantum emitters, QD has the advantage that their material and structures make them compatible with technical solutions in modern optoelectronics. For example, epitaxial processes are used to grow high-quality semiconductor structures with QD [37].

Photonic cluster states are helpful in many quantum information processing applications, including quantum computing and quantum communication. For example, the paper [38] proposes a protocol to generate multidimensional photonic cluster states using a single atom-cavity system and time-delay feedback. The results exist in the heterogeneous integration of different QD with photonic material using wafer bonding [39].

Three dimensions are necessary to realise a topologically fault-tolerant quantum particle cluster state in the required photonic material for 3D holographic memory for PAI. Therefore, multi-dimensional cluster states need to be generated using QD approaches.

However, there are some issues with creating the required material. The problems with the analytical description of multi-photon resonances and their control by external influence have remained. Quantum memories and coherent spin-photon interfaces in QD must store a quantum state for a long time to rewrite and read later. Such interfaces must interact with the environment and be decoupled from it to avoid decoherence of the stored quantum state.

The help of QD studies for creating photonic materials to ensure the long-term storing of quantum particle groups in a fixed structure for coding, writing, rewriting, and reading non-digital images in a single 3D point under laser beams has yet to be made clear.

9.5   Inverse-Engineering

The synthesis of photonic material for PAI is the inverse problem-solving process. Functional requirements and limitations (see also Chap. 5), adequate data sets are amenable to the task input; the output is the structure of the required material and the recommended technology for its production. A genetic algorithm can help to solve such a problem (Fig. 9.1). This algorithm consists of the following steps: building a solution template, making sequential and repetitive implementation of operations (generation of the initial population, reproduction, crossing-over, and mutation), and obtaining a candidate structure of the required material.

The genetic algorithm is advisable to use in conjunction with other classical methods and tools, such as follows:

- Global optimisation and data mining.
- Universal Structure Predictor: Evolutionary Xtallography.

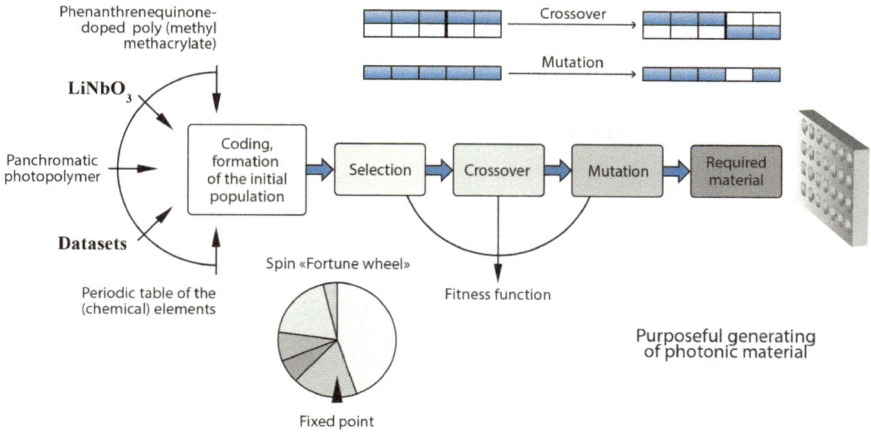

Fig. 9.1 Genetic algorithm for the synthesis of photonic material

- Python materials genomics package.
- DFT.

9.6 Chapter Conclusion

- A unique material for 3D holographic memory must be created—this is the main challenge for PAI construction.
- The requirements for photonic material are achievable by integrating AI and physical-chemistry methods.
- Graphene is a promising matter for synthesising materials and creating a holographic storage device.
- The QD's help in creating photonic materials that store a group of quantum particles in a fixed structure for coding, writing, rewriting, and reading images in local points under laser beams is promising.
- The synthesis of photonic material is a solution to the inverse problem, for which it is worth using a genetic algorithm.

References

1. Oganov, A.R., Pickard, C.J., Zhu, Q., et al.: Structure prediction drives materials discovery. Nat. Rev. Mater. **4**, 331–348 (2019). https://doi.org/10.1038/s41578-019-0101-8
2. Allahyari, Z, Oganov, A.R.: Coevolutionary search for optimal materials in the space of all possible compounds (2020). https://doi.org/10.48550/arXiv.1807.00854

3. Lyakhov, A.O., Oganov, A.R., Stokes, H.T., Zhu, Q.: New developments in evolutionary structure prediction algorithm USPEX. Comput. Phys. Commun. **184**, 1172–1182 (2013). https://doi.org/10.1016/j.cpc.2012.12.009

4. Ringwood, A.E.: The principles governing trace element distribution during magmatic crystallization Part I: The influence of electronegativity. Geochim. Cosmochim. Acta **7**, 189–202 (1955)

5. Vayalamkuzhi, P., Bhargab, D., Joby, J., Rani, J., Krishnapillai, S., Cheranellore, S.K.: High efficiency panchromatic photopolymer recording material for holographic data storage systems. Opt. Mater. **52**, 212–218 (2016). https://doi.org/10.1016/j.optmat.2015.12.042

6. Altshuler, S., Frantz, L.M., Braunstein, R.: Reflection of atoms from standing light waves. Phys. Rev. Lett. **17**, 231–232 (1966)

7. Béguin, A., Rodzinka, T., Vigué, J., Allard, B., Gauguet, A. Characterization of an atom interferometer in the quasi-Bragg regime (2022). https://doi.org/10.48550/arXiv.2112.03086

8. Gamboa, J., et al.: Ultrafast image retrieval from a holographic memory disc for high-speed operation of a shift, scale, and rotation invariant target recognition system. In: Advanced Maui Optical and Space Surveillance Technologies Conference (AMOS) (2022). https://doi.org/10.48550/arXiv.2211.03881

9. Monjur, M.S., Tseng, S., Fouda, M.F., Shahriar, S.M.: Experimental demonstration of the hybrid optoelectronic correlator for target recognition. Appl. Opt. **56**, 2754–2759 (2017)

10. Gamboa, J., Fouda, M., Shahriar, S.M.: Demonstration of shift, scale, and rotation invariant target recognition using the hybrid opto-electronic correlator. Opt. Express **27**, 16507 (2019)

11. Huei, S., Cho, S., Lin, J., Hsu, K.Y., Chi, S.: Influence of fabrication conditions on characteristics of phenanthrenequinone-doped poly (methyl methacrylate) photopolymer for holographic memory. Opt. Commun. **320**, 145–150 (2014)

12. Wang, J., Luo, S.: Volume holographic storage using sync-angular multiplexing by rotating material in thick photopolymer. Opt. Laser Technol. **153**, 108295 (2022)

13. Alcaraz, P.E., Nero, G., Blanche, P.-A.: Bandwidth optimization for the Advanced Volume Holographic Filter. Opt. Express **30**(1), 576–587 (2022). https://doi.org/10.1364/OE.444101

14. Gamboa, J., Hamidfar, T., Vonckx, J., Fouda, M., Shahriar, S.: Thick PQ:PMMA transmission holograms for free space optical communication via wavelength division multiplexing. Appl. Opt. **60**, 8851–8857 (2021)

15. Ko, J.H., Yoo, Y.J., Lee, Y., et al.: A review of tunable photonics: optically active materials and applications from visible to terahertz. iScience **25**, 104727 (2022). https://doi.org/10.1016/j.isci.2022.104727

16. Yang, Y., Yoon, G., Park, S., Namgung, S.D., Badloe, T., Nam, K.T., Rho, J.: Revealing structural disorder in hydrogenated amorphous silicon for a low-loss photonic platform at visible frequencies. Adv. Mater. **33**, 2005893 (2021). https://doi.org/10.1002/adma.202005893

17. Wallace, P.R.: The band theory of graphite. Phys. Rev. **71**, 622 (1947). https://doi.org/10.1103/phys.rev.71.622

18. Novoselov, K.S., Gein, A.K., Morozov, S.V., Jiang, D., Zhang, Y., Dubonos, S.V., Grigorieva, I.V., Firsov, A.A.: Electric field effect in atomically thin carbon films. Science **306**(5696), 666–669 (2004). https://doi.org/10.1126/science.1102896

19. Gein, A.K.: Graphene: status and prospects. Science **324**, 1530–1534 (2009). https://doi.org/10.1126/science.1158877

20. Powel, D.: Silicene: It could be the new graphene. Single-layer sheets of silicon might have electronic applications. Science News (2011). https://www.sciencenews.org/article/silicene-it-could-be-new-graphene. Accessed 10 October 2023

21. Guo, M., Yang, Y., Leng, Y., et al.: Edge dominated electronic properties of MoS_2/ graphene hybrid 2D materials: edge state, electron coupling and work function. J. Mater. Chem. C **5**, 4845–4851 (2017). https://doi.org/10.1039/c7tc00816c

22. Yang, X., Wang, J., Zheng, J., Guo, M., Zhang, R.-Z.: Screening for planar carbon allotropes using structure space sampling. J. Phys. Chem. C **124**, 6379–6384 (2020). https://doi.org/10.1021/acs.jpcc.9b10778

23. Klein, O.: Die reflexion von elektronen an einem potentialsprung nach der relativistischen dynamik von Dirac. Z. Phys. **53**, 157–165 (1929)

24. Katsnelson, M.I., Novoselov, K.S., Geim, A.K.: Chiral tunneling and the Klein paradox in graphene (2006). https://doi.org/10.48550/arXiv.cond-mat/0604323

25. Krekora, P., Su, Q., Grobe, R.: Klein paradox in spatial and temporal resolution. Phys. Rev. Lett. **92**, 040406 (2004)

26. Geim, A., Novoselov, K.: The rise of graphene. Nat. Mater. **6**, 183–191 (2007). https://doi.org/10.1038/nmat1849

27. Ju, L., et al.: Graphen plasmonics for turnable terahertz matamaterials. Nanotechnology **6**, 630–634 (2011). https://doi.org/10.1038/nnano.2011.146

28. Cao, Y., Fatemi, V., Fang, S., Watanabe, K., Taniguchi, T., Kaxiras, E., Jarillo-Herrero, P.: Unconventional superconductivity in magic-angle graphene superlattices. Nature **556**(7699), 43–50 (2018). https://doi.org/10.1038/nature26160

29. Datta, I., et al.: 2D material platform for overcoming the amplitude-phase tradeoff in ring modulators (2022). https://doi.org/10.48550/arXiv.2209.08332

30. Bastrakova, M.V., Munyaev, V.O.: Dissipation entanglement control of two coupled qubits via strong driving fields (2023). https://doi.org/10.48550/arXiv.2310.20229

31. Santos, A.C., Cidrim, A., Villas-Boas, C.J., Kaiser, R., Bachelard, R.: Generating long-lived entangled states with free-space collective spontaneous emission (2021). https://doi.org/10.48550/arXiv.2110.15033

32. Antón, M.A.: Phonon-assisted entanglement between two quantum dots coupled to a plasmonic nanocavity. Opt. Commun. **508**, 127811 (2022). https://doi.org/10.1016/j.optcom.2021.127811

33. Gonzalo, I., Anton, M.A.: Entangling non-planar molecules via inversion doublet transition with negligible spontaneous emission. Phys. Chem. Chem. Phys. **20**, 10523 (2019). https://doi.org/10.1039/C8CP07764A

34. Heindel, T., Kim, J.-H., Gregersen, N., Rastelli, A., Reitzenstein S.: Quantum dots for photonic quantum information technology, p. 113 (2023). https://doi.org/10.48550/arXiv.2309.04229

35. Madsen, L.S., Laudenbach, F., Askarani, M.F., et al.: Quantum computational advantage with a programmable photonic processor. Nature **606**, 75–81 (2022)

36. Michler, P., Kiraz, A., Becher, C.: A quantum dot single-photon turnstile device. Science **290**, 2282–2285 (2000)

37. Pohl, U.W.: Epitaxy of semiconductors: introduction to physical principles. Springer, Berlin Heidelberg (2013)

38. Shi, Y., Waks, E.: Deterministic generation of multidimensional photonic cluster states using time-delay feedback. Phys. Rev. A **104**, 013703 (2021). https://doi.org/10.1103/PhysRevA.104.013703

39. Davanco, M., Liu, J., Sapienza, L., et al.: Heterogeneous integration for on-chip quantum photonic circuits with single quantum dot devices. Nat. Commun. **8**, 889 (2017). https://doi.org/10.1038/s41467-017-00987-6

Chapter 10
Photonic Learning

Training artificial neural networks can be discrete (digital) and continuous (analogue). In the former case, machine learning has always been the sphere of search for ways to optimise this process because it takes enormous amount of computer energy and time. Due to the photonic approach, the latter case eliminates these restrictions and enriches the cognitive semantics of AI models.

10.1 Digital and Analogue Training

Discrete (digital) training of artificial neural networks involves nonlinear digital data processing in multiple layers and can be carried out under or without supervision. In deep learning, optimisation based on stochastic gradient descent is often used. The optimisers usually contain a parameter that indicates the amount of movement of the weights in the direction opposite to the gradient for some small batch of the training sample. If the learning rate is low, it is more reliable but takes a lot of time; if the speed is high, it may not converge. Digital learning can take hours and even days, including on supercomputers. It depends on factors such as

- size and complexity of the data set for training;
- requirements for neural networks' working accuracy;
- neural network architecture and the number of training parameters;
- available computing and energy resources;
- training algorithm and the number of training epochs.

The above applies to traditional approaches to constructing AI systems by representing a continuous (analogue) signal in digital (bits, bytes) form, values in sampling points, counts, and pixels. The digital representation of such a signal cuts down its frequency spectrum, and consequently, the cognitive semantics of AI models are impoverished [1]. The presentation of data by points forces many operations to be

A. Raikov, *Photonic Artificial Intelligence*,
SpringerBriefs in Computational Intelligence,
https://doi.org/10.1007/978-981-97-1291-5_10

performed. For example, producing thousands of operations is necessary to compare two pixel images represented by 1000 points.

The classics of discrete computing cover almost the entire scientific literature pool devoted to AI and machine learning. For example, the "curse of dimensionality" effect is well known: the number of possible configurations of a neural network is usually much greater than the number of training examples. And this curse increases exponentially with an increasing number of parameters. The need for an extensive training sample size makes the training of the neural network unreliable; that is, the neural network will not be able to extrapolate the values of unfamiliar configurations accurately enough. This can lead to incorrect diagnostics of diseases of people, wrong decision-making in transport management, etc.

A discrete neural AI system learns well under the a priori assumption that the trained function can be expressed explicitly as a probability distribution of model parameters whose values are constant and smooth in some local regions. This means the extrapolated function can only change slightly in a small space. However, such assumptions are only sometimes well fulfilled and are scaled in natural conditions with a statistical description of the dynamics of the studied objects. Synthesising artificial datasets is often necessary to train neural networks when working with small datasets. By creating additional data, we can improve the accuracy and robustness of AI systems.

The listed limitations of the discrete data representation and the complexity of training a digital neural network can be overcome by processing a continuous (analogue) signal without preliminary digital sampling. Light, photons, and optics can help to realise the analogue learning path. Say a million floating-point transform operations, required for the Fourier transform in digital case, can be replaced by one operation based on transforming coherent light beams passing through a small optical system. In this case, the signal spectrum is not truncated, and a complete (holistic) signal is used in data processing and training. The operation is performed literally "at the speed of light".

An optical neural network can be built and trained by two approaches:

(a)   Copying the established methods of digital data processing.
(b)   Using all-optical methods of continuous data conversion.

In the (a) case, the optical neural network includes a coherent radiation source (laser), a light beam expander (splitter), holographic lens arrays for filtering light intensity, optical deflectors for directing beams to the required points of the matrix of holographic image storage, the actual matrices of holographic image storage, and a digital controller (computer). The latter must control the discrete optical neural network processes with a conventional discrete computer.

It is challenging to expect multiple accelerations of computational processes from such a digital optical approach, that is, an approach in which optical processes are similar to digital algorithms. The discrete optical data processing paradigm is followed by cutting the spectrum of a continuous signal and, most importantly, excludes the effective representation of non-formalised cognitive semantics of AI models. For such a representation, as shown in our work [1], continuous data from an analogue source needs to be processed without replacing them with values at separated points. The Kotelnikov–Nyquist–Shannon sampling theorem states that such a replacement can be made arbitrarily accurately. However, the discreteness remains in any case; the signal spectrum is reduced, and increasing accuracy requires increasing the number of computational operations.

With a digital optical approach, a neural network's inference time can be significantly reduced compared to its response on a discrete semiconductor computer. A traditional digital supercomputer can implement the output process asynchronously, that is, do parallel calculations in various blocks, which allows for accelerated computations [2]. However, launching a software application that initialises the GPU itself may take a long delay time. This time can be up to some seconds due to the need to clean up the error correction code.

In the (b) case, an optical analogue processor can perform inference and learning operations during the passage of light at a distance of, say, 1 cm; that is, the processing time of a discrete optical network can be around 10^{-10} s. However, this inference speed does not detract from the limitations of training a neural network in traditional, discrete ways.

Can we somehow eliminate the time and energy resources by changing traditional digital optical learning processes?

One way to answer this question is to simultaneously convolution of some optical images in one local point of holographic memory. Optical methods allow some subject images (text, graphics, pictures) to be written into one point (location) of a holographic storage material by superimposing the subject (modulated by the image) and reference beams (Fig. 10.1).

The problem is to write many images in one location by changing the angle of inclination of the object beam and reference beam for every image. A mathematical description of the operation of such a holographic system is quite complex.

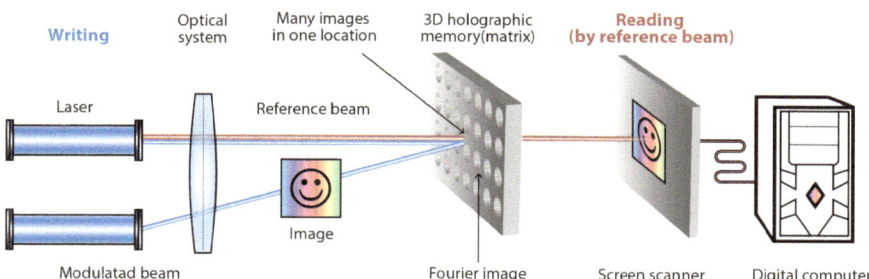

Fig. 10.1 Operation in the holographic storage

10.2 Double Optical Fourier Convolution

To demonstrate the mathematical complexity of writing many images in one location, it is sufficient to show the diffraction integral for the one-dimensional case (x-axis) obtained using Fourier optics methods and reflecting the representation of the light wave formed behind the transparent modulating screen (z-axis) with the recorded image [3]:

$$U_z(x) = \frac{e^{\frac{jkz}{2}}}{\sqrt{j\lambda z}} \int\limits_0^\infty U_0(\xi) \exp\left[j\frac{\pi}{\lambda}(x-\xi)^2\right] d\xi. \tag{10.1}$$

where $U_z(x)$ is the complex amplitude of the object beam at a point x at a distance z from the modulating screen, k is the wave number, $j = \sqrt{-1}$, and λ is the wavelength. Superimposing the object beam on the reference beam incident at certain angles allows the recording of the image on the holographic data storage.

We can try to superimpose several images in one location (point), each creating its diffraction pattern. In that case, the mathematical modelling of Fourier convolution and holographic writing of many images at one point considering different photonic material characteristics can be based on the following reasoning. The convolution of two functions $f(x)$ and $s(x)$:

$$f(x) \otimes s(x) = \int\limits_{-\infty}^\infty f(\xi)s(x-\xi)d\xi = \int\limits_{-\infty}^\infty f(x-\xi)s(\xi)d\xi \tag{10.2}$$

which is known as the convolution or superposition integral, can be realised by multiplication operation and Fourier transform based on Borel's convolution theorem:

$$\mathscr{F}[f(x)]F[s(x)] = \mathscr{F}[f(x) \otimes s(x)], \tag{10.3}$$

$$\mathscr{F}[f(x)s(x)] = \mathscr{F}[f(x)] \otimes F[s(x)], \tag{10.4}$$

where $\mathscr{F}[\varphi(x)]$—Fourier-image of function $\varphi(x)$:

$$\mathscr{F}[\varphi(x)] = \int\limits_{-\infty}^\infty \varphi(x)\,e^{-i\omega x}dx. \tag{10.5}$$

It is followed from the Eqs. (10.3)–(10.4) that for getting the convolution of two functions, it is required to make Fourier-transform operations for every of two functions, multiply the results with getting $G(\omega)$, then make an inverse Fourier-transform of the $G(\omega)$:

$$g(x) = \mathscr{F}^{-1}[G(\omega)] = \frac{1}{2\pi} \int\limits_{-\infty}^{\infty} G(\omega)e^{i\omega x}\,\mathrm{d}\omega. \qquad (10.6)$$

The convolution is a linear operation and is commutative, associative, and distributive, from which it follows that the convolution of a linear combination of a finite number of functions using the addition, multiplication, and Fourier transform operations can be made. That is if we have the finite ($n < N$, where N is finite) set of functions:

$$f_1(x),\; f_2(x), \ldots, f_n(x), \qquad (10.7)$$

for getting the convolution $c(x)$ of functions from the set (10.6), it is required to make Fourier-transform operations for every of these functions:

$$\mathscr{F}[f_1(x),\,]\,\mathscr{F}[f_2(x)], \ldots, \mathscr{F}[f_n(x)], \qquad (10.8)$$

make the multiplication of Fourier images from the set (10.7):

$$\mathfrak{C}(x) = \mathscr{F}[f_1(x)] \otimes \mathscr{F}[f_2(x)] \otimes \cdots \otimes \mathscr{F}[f_n(x)], \qquad (10.9)$$

then make the inverse Fourier-transform of the $\mathfrak{C}(x)$ from (10.9) using formula (10.6):

$$c(x) = \mathscr{F}^{-1}[\mathfrak{C}(x)] = \frac{1}{2\pi} \int\limits_{-\infty}^{\infty} \mathfrak{C}(\omega) \cdot e^{i\omega x}\,\mathrm{d}\omega. \qquad (10.10)$$

This order of getting convolution for the set of functions (10.7) can be followed with the results of convolutions (10.10), such as $c(x)$, as initial functions. The set of such functions—$c_1(x)$, $c_2(x)$, ..., $c_n(x)$—can be the object of application rules (10.8)–(10.10).

The processes of (10.8) and, after they finish, processes of (10.9) can be realised with continuous (non-digital) signals in parallel by an optical device for laser beam transformations. Such a double convolution of finite numbers of the sets of images is required for training PAI's holographic memory.

A double convolution of many functions (images) in an optical system will form a single dual Fourier function (image). Optical multiplication of several images (functions) for realising (10.9) is usually carried out by sequentially passing a laser beam through modulating screens, each carrying one of the images [4]. For the PAI training process, the multiplication of images of one matrix can be replaced by their simultaneous Fourier transform with overlay (adding) into a single point without making the inverse Fourier transform and following recording this result into one of the 3D holographic memory's cells.

For example, an initial set of images may be the analogue photos of facial expressions distributed between different emotional characteristics: anger, happiness, etc.

In the first step, every image has to be Fourier transformed and written by holographic technology into its cell of memory matrix devoted to one of the emotion characteristics. In the second step, such a memory matrix with many Fourier images is used as a modulator of many parallel laser beams for making operations (10.9) and writing the result into one cell of the general matrix. Let us take a look at it in the context of experimental foundations.

10.3 Experimental Foundations

Experimental studies of the double optical Fourier convolution possibility are required—for this, we, together with the "Nantong Triopto PIC Technologies Co. Ltd" company [5], have begun assembling an experimental optical setup (Fig. 10.2).

The experimental study has to include writing a matrix of the modulating screen with many separated images, the Fourier convolution of many images into a single point of the 3D holographic memory, using a matrix of many (some hundreds or

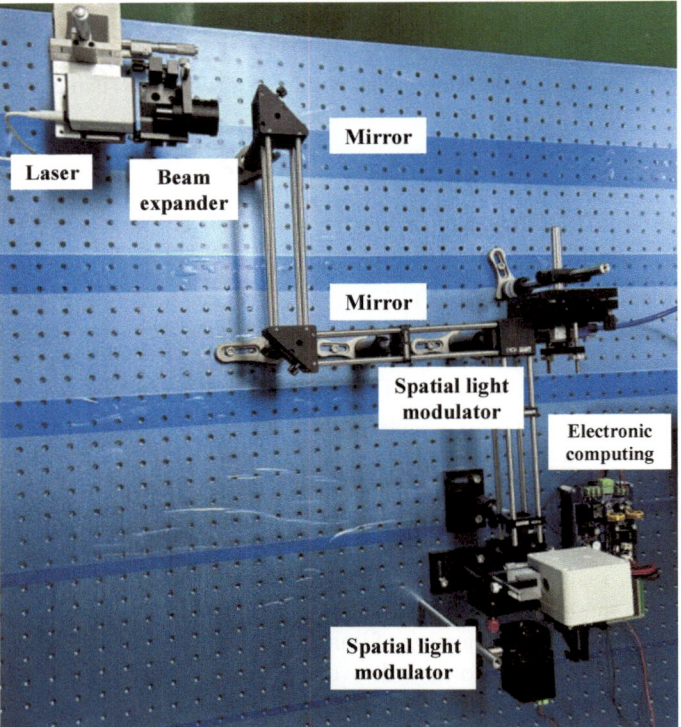

Fig. 10.2 Optical setup with a laser beam path for experiments (published with the permission of "Nantong Triopto PIC Technologies Co. Ltd" company [5])

thousands) microlasers instead of the beam expander (splitter), and optical machine learning processes. The wavelength of lasers may be 405, 488, 532, or 660 nm—the shorter the wavelength, the less space should be used to record the image. Different synthesised photonic materials can be used; however, lithium niobate can be used from the beginning (see Chap. 9).

The idea of constructing the machine learning process of an all-continuous (totally analogue) optical neural network may consist of providing optical convolution of a big set, for example, up to 100, of Fourier images into a single point of the memory. Each point has to be a 3D recording of the big set of images after their Fourier convolution. Such an analogue optical approach will allow:

- instantly modify the architecture, train, and retrain the optical neural network;
- reduce a digital neural network into a matrix of 3D holographic points.

There are examples of trying to implement specific separated components of the photonic AI systems. In the context of such systems, linear optical tools have used diffractive materials [6], spatial light modulators [7], ring resonators [8], wavelength multiplexing techniques [9], and the Mach-Zender interferometers. However, implementations of the nonlinearities and the signal continuity in creating optical neural networks remain challenging; currently, they include digital data representation for network training and inference [10, 11].

The mentioned earlier all-analogue chip combining electronic and light computing (ACCEL) has shown a very high systemic energy efficiency and computing speed [12]. The light-induced photocurrents are directly used for further calculation in an integrated analogue computing chip without requiring analogue-to-digital converters. However, the process of ACCEL's adaptive training resembles an iterative procedure of learning a digital neural network while ensuring the convergence of the weights setting. The number of classes for object recognition is fixed and limited.

The paper [13] addresses creating an intelligent meta-imager architecture for working with a digital back-end by using high-speed and low-power optics for convolutional accelerators for incoherent light. It enables angle and polarisation multiplexing with multiple information channels that perform convolution operations in a single shot. This system achieves 98.6% accuracy in classifying handwritten digits and 88.8% accuracy in classifying fashion images, as shown by experiments. The training database contains 60,000 28×28 pixel images used to train the digital neural network. The channel number was set to 12, and the kernel size for convolution was 7×7. An additional loss function was introduced. All the kernel values are normalised to $[-1,1]$. The training process includes over 50 epochs. The algorithm was programmed using graphics cards based on PyTorch and Compute Unified Device Architecture. This system is based on copying digital algorithms, and the training database is digital. However, the idea of multi-channel architecture may be used to create a non-digital PAI system.

The paper [14] addresses the neuromorphic computer hardware system and convolutional neural networks by exploiting the Fourier-transform method and suitable designed optical tools based on free-space spatial light modulators. This system realised linear and nonlinear operations by optically. The three-layer optical convolutional neural network is presented where the optical part is realised via a caesium atomic vapour cell. This system classifies the handwritten digital dataset [15] with about 84% accuracy. It shows a potential benefit of optical neural networks with nonlinearity. However, this system includes a digital block that transforms analogue signals into digital ones. This digital block connects the nonlinear activated feature maps and the output layer, cutting the signal's spectrum.

Diffractive optical neural networks [16, 17] proposed to train neural connections using the light diffraction effect. For example, papers [18, 19] address a computing architecture where the authors created such a unit using an on-chip diffraction approach and structural re-parameterization. This system was tested for classification with an accuracy of about 91%. The author also builds an optical denoising neural network for regression to handle image Gaussian noise. The paper shows very low energy consumption and high information density.

The operation principles of all the papers mentioned above are based on digital algorithms, which require transforming continuous (analogue) data into digital, followed by signal spectra reduction and losing the power of cognitive semantics of AI models [1]. The problem of creating optical tools for AI systems is also the signals' optical distortions, including aberrations, which reduce the effect and accuracy of the transformations. In addition to the interference of signals, numerous theoretical studies show that to present analytically (formulas) the interference process of adding several hundred rays, each carrying information about a particular image, is excessively cumbersome, and obtaining a sufficiently accurate result is very challenging.

10.4 Instant Learning

Instant optical learning can be realised by converging "bottom-up" and "top-down" processes. The former comes from the subatomic level, the letter—from the idea of the whole.

The classical formation of the whole from parts, the "bottom-up" approach, splits the whole down to the level of quantum particles, limiting its study's completeness and universality. In quantum terms, symmetry is understood as the transformation of fields in which the function of the action does not change. The concept of action density or Lagrangian is used to define the action function, which depends on the value of the field and its derivatives specified in the coordinates of spatial variables and time. Unfortunately, classical theory and the Standard Model of quantum physics cannot fully explain the mass spectrum of elementary particles and combinations (groups) of their structures. The action of such a group leads to different mass levels within the charge multiplets using reduction. It is described by the equation considering the charge and hypercharge, parity and charge-parity symmetries, spin, and isotopic spins of the particles [20, 21]. In the quark representation of the quantum model of the human brain and Artificial Mind (see Chap. 2), the spin of an elementary particle is equal to the total moment of the amount of motion of the corresponding moment of the quark system and the total spin of the quarks. However, spin is not a mechanical concept. This classical representation of symmetry in the quark model faces great difficulties when describing the excited baryons spectrum.

In the Standard Model, where the "fundamental" quarks, leptons and bosons are located, the number of basic parameters is 18, 13 of which are related to the masses of fermions and quarks. However, more than this number of parameters are required in practice. For example, detecting neutrino oscillations involves expanding this model by introducing new masses, the parameters of which have only been determined experimentally so far. The reason for the incompleteness of quantum physics, which has been preserved for decades, is probably the reliance on the classical system approach, namely the reductionist atomic hypothesis, the essence of which is that the properties of an integral particle system are entirely and by the principle of separability determined by the states of its subsystems. This required introducing parameters such as Planck's constant and the speed of light, which set the limits of applicability of classical physics.

To create PAI, including synthesising the required photonic material and developing a method for rapid learning of PAI, it is necessary to do precisely the opposite—to go from the whole, e.g. "top-down". After all, it is being confirmed with increasing certainty that parts of a quantum system are in a non-separable (entangled, coupled, superposed) state, and the properties of a system taken as a whole are far from being entirely determined by the properties of its parts. That is, each quantum subsystem does not have its pure state. This holistic approach does not have to be tied to the idea of space–time. The initial attempt to construct such a holistic approach was Heisenberg's work on the nonlinear spinor theory of matter, in which he introduced the concept of an elementary (universal) length commensurate with the classical radius of the electron and considered the observed particles as excited states of the field [22]. In this case, the characteristic scale of the rest masses of the particles is set using this universal length. Further work in this direction can allow for some reduced formalisms under the construction of the mass spectrum of elementary particles [23, 24].

The purposefulness of creating PAI learning algorithms is supported by ideas from this non-classical holistic approach to quantum physics, which departs from the quark model based on flavour symmetry groups or local gauge symmetry with their breaking [25]. To date, several assumptions have been made regarding the essence of an elementary particle and the possibility of its structuring [26]. For example, a particle is a collapsed wave function (reduction of quantum superposition), quantum excitation of a field, irreducible representation of a symmetry group, many layers system, deformation of qubits, and so on. That is, the concept of an "elementary particle" is reduced to a specific shade of a particular substance (whole), and a particle (quantum) is a derivative of the field phenomenon.

Such a view suggests that a system for storing multiple holographic images at individual points and training the analogue optical neural network should be built on holistic principles, considering the non-trivial structure of elementary particles. At the same time, particles should be regarded not only traditionally as both particles and waves but interpreted as a kind of emergent construction generated by the perturbation of the whole (energy, substance) and being the antithesis of the idea of a space–time continuous continuum that provides the transfer of interaction from point to point with a limited speed. Then, the "spectrum of elementary particles" is replaced by the spectrum of integral matter (substance). Such a view allows for an analytical description and practical implementation of fast learning processes in PAI systems. Therefore, the study of the phenomenological component in the design of an optical computer and the need for appropriate experiments in this design are critical.

Considering the above, the training system of an analogue optical neural network can be represented in Fig. 10.3. This system provides PAI learning in two steps. In the first step, an optical matrix of training images—the transparent training matrix, is formed in Subsystem 1. The training matrix is recorded by writing a big set (several hundred) of training images (pictures, graphics, texts, etc.) in its cells. The holographic method ensures the ability to write one image in one cell using a light modulator to create subject (modulated) beams. The subject beam and the reference beam of lasers carry out the recording. The direction of recording beams is carried out with the help of deflectors, which are controlled by a digital system.

Subsystem 2 is used for training PAI by the one step. After filling the training matrix by Subsystem 1, the big set of reference beams is directed to this matrix using a laser beam expander (splitter) for parallel reading of the recorded training images. This system simultaneously makes Fourier convolution of all images recorded in the training matrix, see (10.10). Using a digital controller and deflectors, the system directs all modulated by the matrix of training image beams to one of the points of

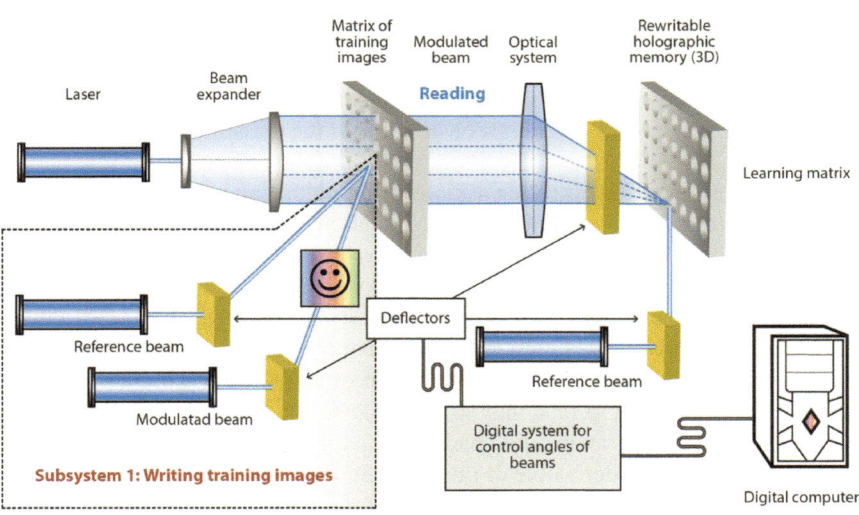

Fig. 10.3 Analogue optical neural network training system

the learning matrix of Subsystem 2, which is the rewritable 3D holographic matrix. Another training matrix is formed to register another point (cell) of this matrix.

With such an optical machine learning system, the direct learning time in the presence of a recorded training matrix will be determined by the time of light passing a distance of several centimetres, unlike modern values of the long-time learning (hours and days) measured for the learning with a digital supercomputer.

Such a learning matrix can be used in the PAI system for image recognition. For example, see Fig. 3.2 (Chap. 3), where the analogue memory matrix of emotions is used for emotion recognition by matching the expanded (split) input analogue image with all cells of the trained learning matrix simultaneously.

10.5   Chapter Conclusion

- The training of artificial neural networks can be discrete (digital) and analogue (continuous).
- The learning rate of digital artificial neural networks can take hours and days on supercomputers.
- Processing a continuous signal without its preliminary digital representation using PAIT can overcome the complexity of training a digital neural network, including time and energy resource requirements.
- The PAI system has to ensure the writing and rewriting of transparent images in the form of a holographic matrix with separate images in its cells.
- The PAI can be designed in such a way as to ensure the processes of training and recognition can be realised almost instantly.
- An experimental optical setup has been assembled to check the possibility of double Fourier convolution and holographic writing of many images in a single point.

References

1. Raikov, A.: Cognitive semantics of artificial intelligence: a new perspective. In: Topics: Computational Intelligence XVII, pp. Springer Singapore (2021). https://doi.org/10.1007/978-981-33-6750-0
2. Deng, Y., Guo, M., Ramos, A.F., et al.: Optimal low-latency network topologies for cluster performance enhancement. J. Supercomput. **76**, 9558–9584 (2020). https://doi.org/10.1007/s11227-020-03216-y
3. Stark, H. (ed.): Application of optical fourier transforms. Academic Press, London (1982)
4. Liu, J., Wu, Q., Sui, X., et al.: Research progress in optical neural networks: theory, applications and developments. PhotoniX **2**, 5 (2021). https://doi.org/10.1186/s43074-021-00026-0
5. Triopto PIC. https://www.triopto.com/. Accessed 25 Dec 2023
6. Lin, X., Rivenson, Y., Yardimci, N.T., Veli, M., Luo, Y., Jarrahi, M., Ozcan, A.: All-optical machine learning using diffractive deep neural networks. Science **361**, 1004–1008 (2018). https://doi.org/10.1126/science.aat8084

7. Zhou, T., Lin, X., Wu, J., Chen, Y., Xie, H., Li, Y., Fan, J., Wu, H., Fang, L., Dai, Q.: Large-scale neuromorphic optoelectronic computing with a reconfigurable diffractive processing unit. Nat. Photonics **15**, 367–373 (2021). https://doi.org/10.1038/s41566-021-00796-w
8. Tait, A.N., Nahmias, M.A., Shastri, B.J., Prucnal, P.R.: Broadcast and weight: an integrated network for scalable photonic spike processing. J. Lightwave Technol. **32**(21), 4029–4041 (2014). https://doi.org/10.1109/JLT.2014.2345652
9. Feldmann, J., Youngblood, N., Wright, C.D., Bhaskaran, H., Pernice, W.H.: All-optical spiking neurosynaptic networks with self-learning capabilities. Nature **569**, 208–214 (2019). https://doi.org/10.48550/arXiv.2102.0936
10. Zuo, Y., Li, B., Zhao, Y., Jiang, Y., Chen, Y.-C., Chen, P., Jo, G.-B., Liu, J., Du, S.: All-optical neural network with nonlinear activation functions. Optica **6**, 1132–1137 (2019). https://doi.org/10.48550/arXiv.1904.10819
11. Ryou, A., Whitehead, J., Zhelyeznyakov, M., Anderson, P., Keskin, C., Bajcsy, M., Majumdar, A.: Free-space optical neural network based on thermal atomic nonlinearity. Photon. Res. **9**, B128–B134 (2021). https://doi.org/10.48550/arXiv.2102.04464
12. Chen, Y., Nazhamaiti, M., Xu, H., et al.: All-analog photoelectronic chip for high-speed vision tasks. Nature **623**, 48–57 (2023). https://doi.org/10.1038/s41586-023-06558-8
13. Zheng, H., et al.: Intelligent Multi-channel Meta-imagers for Accelerating Machine Vision (2023). https://doi.org/10.48550/arXiv.2306.07365
14. Yang, M., et al.: Optical convolutional neural network with atomic nonlinearityю (2023). https://doi.org/10.48550/arXiv.2301.09994
15. LeCun, Y.: The mnist database of handwritten digits. http://yann.lecun.com/exdb/mnist/ (1998). Accessed 15 Nov 2023
16. Xu, Z., Yuan, X., Zhou, T., et al.: A multichannel optical computing architecture for advanced machine vision. Light Sci. Appl. **11**, 255 (2022). https://doi.org/10.1038/s41377-022-00945-y
17. Yan, T., et al.: All-optical graph representation learning using integrated diffractive photonic computing units (2022). https://doi.org/10.48550/arXiv.2204.10978
18. Huang, Y., Fu, T., Huang, H., Yang, S., Chen, H.: Sophisticated deep learning with on-chip optical diffractive tensor processing (2022). https://doi.org/10.48550/arXiv.2212.09975
19. Fu, T., Zang, Y., Huang, Y., et al.: Photonic machine learning with on-chip diffractive optics. Nat. Commun. **14**, 70 (2023). https://doi.org/10.1038/s41467-022-35772-7
20. Okubo, S., Ryan, C.: Quadratic mass formula in SU 3. Nuovo Cim **34**, 776–779 (1964). https://doi.org/10.1007/BF02750019
21. Beg, M., Singh, V.: Splitting of the 70-Plet of SU(6). Phys. Rev. Lett. **13**, 509–511 (1964). https://doi.org/10.1103/PhysRevLett.13.509
22. Heisenberg, W.: On the mathematical frame of the theory of elementary particles. In: Blum, W., Dürr, H.P., Rechenberg, H. (eds.) Original Scientific papers. Wissenschaftliche Originalarbeiten. Gesammelte Werke. Collected Works, vol. A/3. Springer, Berlin, Heidelberg, pp. 158–165 (1993). https://doi.org/10.1007/978-3-642-70079-8_13
23. Koide, Y.: New view of quark and lepton mass hierarchy. Phys. Rev. D. **28**. 252 (1983). https://doi.org/10.1103/PhysRevD.28.252
24. Barut, A.O.: Lepton mass formula. Phys. Rev. Lett. **42**, 1251 (1979). https://doi.org/10.1103/PhysRevLett.42.1251
25. Bernabeu, J.: Symmetries and their breaking in the fundamental laws of physics, p. 29 (2020). https://doi.org/10.48550/arXiv.2006.13996
26. Wolchover, N.: What is a Particle? Quantamagazine. **12**. (2020). https://www.quantamagazine.org/what-is-a-particle-20201112/. Accessed 25 Dec 2023

Conclusion

The digital way of developing AI systems has ancient origins and experience of binary descriptions of nature. In the computer age, this became convenient for computer creation. However, with the increasing computer performance, the digital base has faced severe limitations, such as the impossibility of further miniaturisation of the semiconductor element and the need for high time and energy costs for machine learning. The digital way cannot fully represent the non-formalisable cognitive semantics of AI models, which also limits the growth of accuracy in their implementations.

Currently, the existing analogue (continuous) alternative of digital AI systems development based on optical (photonic) methods, as usual, includes elements that realise digital computer algorithms with the requirement to sample continuous signals. The optical way provides a significant energy performance effect; however, in the case of reliance on digital algorithms, the further development of optical AI systems is accompanied by almost the same limitations as in the case of the implementation of AI systems on a digital basis.

This book is about the total optical (photonic) way of AI systems development, which can eliminate the limitations of digital computing and algorithms. The photonic way opens up new options for considering AI models' non-formalisable cognitive semantics, manifesting themselves when AI systems' instrumental basis is immersed at the subatomic level and begins to consider environment space and quantum non-local effects of human consciousness.

The creation of Photonic AI (PAI) systems began many years ago, but significant challenges have given way to the development of digital systems. The optical way required new math for optical transforms, new material for rewritable holographic memory, and a particular digital controller must be developed.

In the last decades, we have witnessed the creation of many different optical tools for calculation support. For AI systems, optical devices can help realise neural networks. Optical instruments can do arithmetic, logical operations, Fourier transforms, and math convolution.

© The Author(s), under exclusive license to Springer Nature Singapore Pte Ltd. 2024
A. Raikov, *Photonic Artificial Intelligence*,
SpringerBriefs in Computational Intelligence,
https://doi.org/10.1007/978-981-97-1291-5

However, optical computing cannot currently replace the entire functionality of digital computing. It can replace only selectively chosen parts of digital applications. Some technological problems, including creating unique photonic materials for rewritable 3D holographic memory and decreasing energy and latency costs of analogue-to-digital conversion for connection with digital computers, limit the development of PAI. The issues of making the required PAI system also are as follows:

- How does the PAI overcome the problems of optical nonlinearity and vulnerability to noises and system errors?
- How does PAI solve inverse problems by considering dynamic characteristics or movement of objects?
- How can PAI better connect with digital, quantum, bio, and neuromorphic computers?

There are problems related to processing sound, taste, and smell. The architectural implementation of these components of human abilities in PAI can be similar—their architecture can be reduced to a universal solution. Device designs for these processes will only differ in data collection methods. The sensors of sound (microphone), taste ("electronic tongue"), and small ("electronic nose") have already been developed and implemented commercially. For example, in the case of sound, air vibrations affect the microphone membrane, which converts sound waves into a continuous signal using an electromagnetic system, such as a condenser. With the photonic implementation of the computer, the signal is not digitised and is processed in its original frequency-amplitude-time form.

The issue remains optical processing and fixing the time dynamics of various fluid processes. A digital computer must introduce a time parameter for such a process. In that case, discreteness is introduced into the data processing, and the process is encoded, thereby losing its natural analogue form. Suppose only analogue signal images are stored on an optical storage device. In that case, it will be necessary to change the direction of the laser beam with the help of a digital controller with time discreteness and sequentially read these images. However, the discrete reading order of analogue images is too long term.

This book tries to overcome these limitations on a theoretical level and suggests some technological and experimental decisions. The conducted research allows us to draw the following conclusions:

- A human neuron system performs functions in analogue and digital ways; the natural analogue signal propagates slowly, flexibly and unreliably; between the neurons, signals are transmitted by pulses at a speed of up to 70 m/s, quite accurately and reliably; the human brain contains a high volume of "dark" neurons that have a specific localisation and a fuzzy boundary, which is associated with the dark matter and dark energy in the Universe—these aspects can be more fully taken into account using optical computing.
- In modern artificial neuron systems, data are processed discretely and reliably at the speed of an electromagnetic wave in a cable; signals are transmitted in binary

(pulse) form; to share data by pulse, an error is introduced into a continuous signal, and its spectrum is distorted (cut off) due to the difference in the values of analogue and interpolated discrete signals—these limitations can be overcome taken into account optical computing.

- The further fundamental development of AI to the Artificial Mind version in the form of PAI for considering deep feelings, thoughts, and transcendental states of mind requires taking into account the analogue nature of natural neurons' behaviour, including quantum non-local and fluctuation effects.

- Using thermodynamic, quantum, and wave approaches helps to represent the human thinking processes that lie much "deeper" than the digital neural network semantic interpretation; for getting synergy, turning to methods for inverse problem-solving to find the necessary conditions for converging PAI processes to goals is required.

- PAI can monitor human emotions continuously and cover thousands of shades of emotions; it is helpful to introduce the term "Photonic psychology", which embraces digital and continuous psychological aspects of creating and using PAI and helps accelerate psychological procedures and strategic planning meetings.

- The main problem of creating PAI consists of synthesising unique materials for 3D holographic memory, which can store many rewritable images in one local point; the requirements for such material are achievable by integrating physical-chemistry methods and inverse problem-solving methods.

- The idea of hypothetically creating the PAI system for copying non-symbolic thought operations may be realised based on laser solitons and 3D rewritable holographic memory to represent and process cognitive semantic interpretations of AI models.

- Natural neurons consist of atoms, and their behaviour depends on their subatomic structure; therefore, the understanding of thinking processes and their representation by PAI should consider the structure of elementary particles and photons' behaviour, which carry information about the source.

- The complexity of training a modern digital neural network can be overcome by optical processing of a continuous signal without its preliminary digital representation; the PAI system provides parallel convolution of a significant amount of training images and can be done almost instantly without requiring substantial energy and time resources.

Glossary

Artificial Mind AI phenomenon which takes into account feelings, thoughts, and transcendental states of mind while considering as discrete as the analogue (continuous) nature of human neurons' behaviour on a subatomic level, including quantum non-locality and fluctuation effects

Dark neurons Hypothetically, inactive neurons of the human brain that can only be activated together and send impulses one or a few times during a person's lifetime

Double optical Fourier convolution A two-step optical Fourier convolution: the first is the convolution and recording set of images into the translucent memory matrix's cells; the second is the optical simultaneous convolution of all images previously recorded into the translucent matrix

Human thoughts Not fully formalised phenomena of human consciousness that can be intrusive and stable, elusive and unstable; unscripted ones quickly disappear and recorded—undergo semantic distortions

Laser solitons Stable structures of coherent waves that move in space and combine into complexes without distorting each other

Optical convolution An optical operation of Fourier convolution of images

Photonic material A material with a structure that a laser beam can change on a subatomic level; the result is stored for a long time and is not altered during reading by laser

Photonic psychology The possible branch of computer psychology, which embraces digital and continuous psychological aspects of creating PAI

Quantum spot (dot, point) A small volume of photonic material that can holographically store for a long time many different and separable states of a group of subatomic particles written and read by laser beam

Rewritable 3D holographic memory Analogue (non-digital) holographic storage that includes data on a photonic material plate in the form of a matrix of images set, where each of the matrix's cells corresponds to a single image or class of images

Analogue AI AI system with a single-layer holographic memory matrix which realises training procedures by a single optical operation which processes natural continuous signals without its previous sampling or digital transform